Do You See What I See?

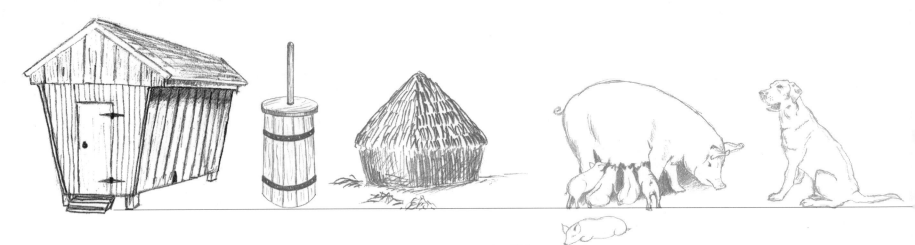

Other books by Jim Ellis:

Blessed By Buzzards
An Old Dog Blessed With New Tricks
Waiting For The Other Shoe To Drop
From Malt To Marriage

Do You See What I See?
Treasured Tales of Life on The Family Farm

Jim Ellis

ORANGE *frazer* PRESS
Wilmington, Ohio

Acknowledgement

During one of our trips west we stopped on a highway superseded by an Interstate in Nebraska and took two pictures. There were no cars to be seen in either direction. I stood looking west and Anne was on the lookout to the east. We could see no beginning and no end to this long stretch of road. Recognition of all who had a part in the production of this book is like that road. Where to begin and where to end is the question. It seems right to begin with gratitude to God for his gift of life and our loving parents who were family farmers. A large measure of thanksgiving goes to the friends who shared their stories of family farm life. Even if the book had not been published the satisfaction of sharing the stories in small groups will not be forgotten. The end is not in sight because of the readers who will be inspired to share their family farm stories with others.

My wife Anne has been there for me and given me a generous share of her computer time. She had major input in the final proofing. In addition to Anne's help, Granddaughter Holly Hardin has once again greatly enhanced the book with her creative art. Gratitude also goes to others who have been generous with their help publishing the book: Marcy Hawley and the staff of Orange Frazer Press, Chad DeBoard, Janice Ellis and Sarah Hawley. Thanks to Dover Friends Meeting and Dan and Marla Stewart of Books "N" More for their encouragement and support at times of signing and sales.

ISBN 978-1933197-777
Copyright©2010 Jim Ellis

No part of this publication may be reproduced in any material form (including photocopying or storing in any medium by electronic means and whether or not transiently or incidentally to some other use of this publication) without the written permission of the copyright holder except in accordance with the provisions of the Copyright, Designs and Patents Act 1988.

Additional copies of *Do You See What I See? Treasured Tales of Life on the Family Farm* may be ordered directly from:

Jimanne Ellis		Orange Frazer Press
402 Prairie Ave.	or	P.O. Box 214
Wilmington, OH 45177		Wilmington, OH 45177

Telephone 1.800.852.9332 for price and shipping information.
Website: *www.orangefrazer.com*

Book and cover design: Chad DeBoard, Orange Frazer Press
Cover photo: Stoltzfus dairy on Morris Road, near Sabina, Ohio. Photograph by Ron Levi: www.ronaldgleviphotography.com
Illustrations: Holly Kathryn Hardin

Library of Congress Cataloging-in-Publication Data

Ellis, Jim, 1927-
  Do you see what I see? : treasured tales of life on the family farm / Jim Ellis.
    p. cm.
  ISBN 978-1-933197-77-7 (alk. paper)
    1. Farm life--United States--Anecdotes. 2. Family farms--United States--Anecdotes. I. Title. II. Title: Treasured tales of life on the family farm.

S521.5.A2E45 2010
630.973--dc22
                                        2010032255

Honoring the memory of our parents
J. Warren and Helen Madden Ellis
Walter A. and Anna Henley Coble
For their gift of life on diversified family farms

J. Warren and Helen Madden Ellis

Walter A. and Anna Henley Coble

Table of Contents

Introduction	xiii
A Walk Around The Farm	1
That's The Way It Is	9
Hold Your Horses	27
Tractors Take Their Turn	47
Let There Be Light	64
Heart of The Home	74
Breeds and Brands	95
Saturday Night—For What It's Worth	120
Helping Hands and Feet	135
The More We Get Together	148
Why Pay to Play?	167
Now You See It—Now You Don't	187
Saved Labor—Lost The Farm	193

Introduction

The title "Do You See What I See," by design calls the reader to look back in time to the years before the pictured farmstead. A farmstead is the cluster of buildings, equipment and structures essential to the successful functioning of the family farm. Those who lived in and around such farmsteads during the years from 1930 to 1950, have no trouble completing the picture with the help of the eyes of memory. The following is a partial list of what could be seen on farmsteads of that time.

Take a closer look with me as we recall the many defining features that were basic to the picture in times past. Barn, house, outbuildings, fields, hog houses, hog boxes, hog feeders, hog water tanks, milk house, silo, straw stack, windmill and stock watering tank, chicken house, chicken brooder house, cattle feeding shed, corncribs, tool shop, woven wire fence with upper strand of barbed wire, wooden and metal gates, pasture fields, fields for corn, oats, beans, wheat and hay, farm house, white picket fence, outhouse, smokehouse, springhouse, wood/coal shed, one car garage, front yard, back yard with clothes line, garden for corn, tomatoes, beans, peas, beets, potato and melon patch, rhubarb and asparagus beds, berry bushes, grape arbor, fruit orchard, woods, and a stream for the fortunate. The completed picture also needs livestock such as: horses, cattle, pigs, sheep, chickens, turkeys, dogs and cats.

Some of these may have been background to the barn pictured on the book cover. Each structure was designed and defined by its operational function for the family farm. However, basic to the stories shared for this book is something of a spiritual dimension. Many of the structures were sanctuaries of soul-searching and satisfaction. While the family farm house was a home and obviously sacred/profane, it would be short sighted not to see the sacred in a picture such as comforted cattle sheltered in the barn on a cold winter night. Add to the picture young lambs, calves and colts kicking up their heels in a frolic on a warm spring morning, as well as pigs wallowing in cool soupy mud in the heat of the summer sun.

Do you see what I see?

The bulk of the material in this book came from friends with memories of life on a family farm in the period between 1930 and 1960. Lest some readers be misled, this writing is not an unbiased objective rural sociological survey. The invitation to interview given to interested persons was as follows:

If you were raised on a diversified family farm in the 1930–1950s or lived in a small town in a farming area, I'd like to meet with you. I grew up on such a farm in Illinois and would like to talk with you about those memorable years.

Living on a family farm with diversification of livestock and grain brings to mind stories that need to be shared. We have lived at a time of dramatic change for farm families and their small town neighbors. Today in many areas livestock and crop diversification along with family farms is non-existent. With your help, I'd like to write stories about life on the farm with all its highs and lows. For me the best of times far overshadowed the worst. I hope it was the same for you and would like to hear your stories.

The stories told me are shared with the hope you too can see and appreciate the basic features of a functioning family farm. Beyond that is the hope you can also see the working relationships of families in rural communities and small towns; families blessed by the spirit of cooperation and gratitude to God for life on the land.

Prior to graduating from high school, I was an active participant in our diversified family farming operation. As of this writing, I am 83 years old. My mind has difficulty encompassing the magnitude of the changes since the 1930s. When we consider the dramatic transition from strenuous hand labor to our present laborsaving hand-held remotes, it is truly amazing! From my vantage point of 80 plus years, 1930–1950 brought the greatest years of rapid change and challenge for agriculture, rural life and the family farm. This book is written against the backdrop of a dwindling number of family farmers who remember these challenging years. I am grateful for the many people who have been willing to share their treasured stories of life on a family farm. The focal point of each story is as told by each person interviewed. Details building up to the actual experience are in part colored by similar happenings for the author. It has been said the "Good Old Days" were at one time referred to as "These Trying Times." In spite of all the sacrifices, the majority of those I interviewed have memories of the family farm as a "wonderful way of life" known to only a few farming families today.

My wife Anne and I were in Shelby, Montana, for breakfast. When we walked out of the restaurant after breakfast, I did not realize I'd forgotten to pick up my cap. We were over fifty miles from Shelby on our way to Great Falls, before I missed it. Even though this cap was a treasured gift, I decided it was not worth the loss of time and one-hundred miles extra driving to go back and retrieve it. However, if it had been a family member or friend, there'd be no question of what to do.

Since the tough times of the 1930s family farmers have progressed from subsisting on farm produce with little financial expense, to major financial expense for equipment and products to make the farm more income producing. Today, some farmers are asking if the dramatic change from high investment of physical labor with low financial investment, to low input of physical labor and high financial investment is really progress. Progress has not been without a price. The stories I've heard confirm my suspicion that we have left some things behind which we would be well served to go back and attempt to recover. Among these things is a wholesome understanding and acceptance of our personal identity based on who we are rather than what we have. Add to that our need to accept each other and work together for the common good. Finally we need to acknowledge our dependence on intimate bonding with our Lord, land and livestock. In years past, we knew our neighbors because we met with them on many levels of community involvement. Family farms, along with small town businesses, rural schools, churches and rural organizations are now seen only in the hazy mist behind collapsing and abandoned barns.

Family farms with a diversity of livestock and grain were not totally dependent upon a single income source. The basic satisfactions of a good life were not limited to financial success. Bonding with people, livestock and the soil was central to life on the family farm. For the most part, working, playing and praying together were gained by the coin of neighborliness and willing participation.

The stories told to me, I share with you because they may be lost if not shared. They are rich in the diversities of life. As you read and enjoy these stories of our rural past, I hope you will pick up the thread of cooperation and community. Triumph and tragedy, humor and sober reflections were all woven into the tapestry of life on the family farms and in neighboring towns.

Consideration is given to the dynamics of the changes in the life style and farming methods of families living in the 1930s through the 1950s. This is a collection of human interest stories of the challenges and satisfactions dealing in part with the adjustment to the changes.

Consideration is also given to the impact of mechanization and changes in laborsaving devices in the home as well as on the farm. This involves speculation concerning the erosion of the diversity of livestock, poultry and grain with its impact on rural community life and the family farm. The time has come for examination of the consequences of the agribusiness thrust for greater production and economic gain on the small family farm and the surrounding rural community.

Do You See What I See?

A Walk Around The Farm
A Tour of The Features of The Farmstead

Then God said, "I've given you every sort of seed-bearing plant on Earth and every kind of fruit-bearing tree, given them to you for food. To all animals and all birds, everything that moves and breathes, I give whatever grows out of the ground for food." And there it was. God looked over everything he had made; and it was so good, so very good!
Genesis 1: 29-31–Eugene H. Peterson *The Message*

Much like a city or large town a self-sustaining, diversified family farm in the 1930s –1950s had divisions of labor and buildings or features necessitated by the farm diversity. On a family farm people and livestock required care, feeding and a place of protection. Produce of grain, hay, meat and milk needed places of storage as did the equipment and instruments of labor. Farmstead features including every structure from the picket fence to the bank barn were guardians of the farm family, livestock, feed and equipment.

During the 1930s and 1940s the farmstead of my youth in East-central Illinois was concentrated in ten acres near a county road on the western edge of the 300-acre family farm. Most of the farmstead features were confined to ten acres with one exception, a house and barn near the center of our land. It was known as the "House in the Fields." In my youth I had little understanding of the significance of the placement of that house and barn. All I knew the unoccupied buildings were rather spooky and down a long dirt lane. My great-grandfather Henry Thornburg Ellis built the house and barn central to the farmstead near the county road. I haven't a clue as to the builder of the "House in the Fields," but I feel certain I know why it was there. The separation between the two sets of buildings is a simple fact of the difference between the power of horses and the power of mechanization. Without roads or even much need of them, putting the farmstead near the center of the farm was the best arrangement. Activity related to farming radiated from the rudimentary central farmstead in all directions. Since most of the necessities for living were raised and processed at the farm site prior to the advent of mechanization and laborsaving devices, there was little need for good roads. When tractors and automobiles replaced the horse and carriage, it became necessary to market farm products to be sold for money to pay for these new laborsaving devices.

Vehicles for taking products to market in nearby towns and villages required better roads and a farmstead nearer the public road. I had the good fortune of living a few of the "horse power" years and through the transition to "tractor power." As I think about it now, I am amazed on our farm at least, that the transition took so little time. We had 4 or 5 teams of draft horses and one steel-wheeled Model D John Deere in the mid 1930s. Three years later we had two tractors and not one horse. For us the transition was complete. The horses were turned out and the tractors turned on.

At this point it seems in order to include a brief description of some features and fixtures of a fairly typical farmstead in the period between 1930 and 1960. An overview of many farms in the Midwest would show the acreage outlined with woven wire fence or a fence row of trees and shrubs. Clusters of buildings and small fenced-in lots circle the primary barn and the farm house. For the readers who have little knowledge of a family farm, I have a brief description of some of the features and fixtures basic to the family and operation of the farm. For those who have lived on a family farm during these targeted years, I trust it will bring to mind mostly fond memories.

Fields

In my mind I walk the fence rows of the farm in Illinois. Today with all the fences ripped out it is no longer the farm that set the boundary lines of my life. Some fields were for animal pasture or hay, other fields for grain production with a different crop every year. Most farm fields had woven wire fencing with a top strand of barbed wire for livestock grazing or gleaning for corn left in the field after harvest. Today the fence rows are gone. It is one large field for grain.

"He only says, 'Good fences make good neighbors.' Spring is the mischief in me, and I wonder if I could put a notion in his head: 'Why do they make good neighbors? Isn't it where there are cows? But here there are no cows. Before I built a wall I'd ask to know what I was walling in or walling out, and to whom I was like to give offense.'" These few lines from Robert Frost's poem "Mending Wall" are grounded in a long standing rural tradition. Farm fences over time do become weak and worn. Certain cows are self-appointed fence inspectors that will find a weak spot and use it as easy access to the greener grass on the other side. "Your cows are out!" is never welcome news to a busy farmer.

Press A. and two brothers were driving a few cows along the roadside on the outside of the farm fence in the early 1930s. Had you come upon this sight at the time, the chances are you would have concluded the cows managed to find a weak spot in the fence. The follow-up conclusion would have been the boys are trying to drive them back in the pasture where they belong. Under normal circumstances you would have assumed correctly.

But this was not one of those "the cows got out," days. This summer in the 1930s was one of the hottest and driest on record. Press and his brothers were herding their cows along the roadside outside their farm fence because it was about the only place around where there was grass left to feed the cows. This day the boys were herding the cows to feed on the grass along the roadside for a mile square near their farm. The boys spent their time sauntering around while the cows walked and ate grass. It was by no means a fast pace. They had time to tease and torment each other. The gravel roads had an abundance of stones that begged to be thrown somewhere. They were careful not to hit the cows, just keep them off the sparsely traveled road. They mostly aimed at telephone poles and an accidental hit of a glass insulator when trying to break the boredom.

They were about halfway around the mile square, when a change in the weather created a bit of excitement. Press and his brothers had taken little notice of the clouds drifting in from the west. The big drops of rain were a welcome sight for the farmers who could watch the torrent from the shelter of a shed or barn. The boys and the cows caught out two miles from home had no shelter from the deluge. The cows with heads lowered lined up along the fence without the slightest inclination to hurry back to the barn. The boys soaked to the skin, knew they dare not abandon the cows. Fortunately there were no lightening strikes nearby.

Around the Barn

Barn: Heart of the farming operation. Functioning barns in the 1930s were built to house livestock and store grain, fodder, hay and straw. Barns were built in a variety of sizes, but with some degree of sameness dictated by their purpose and use. Many barns were three story "bank barns" built on the side of a hill. Livestock was usually housed on the ground floor. The second floor, with a hillside inclined drive entrance was used to store grain and equipment. Hay and fodder stored on the top floor were brought in through a large open space in the middle floor, or an outside opening near the very top of the barn.

Cattle Feeding Sheds: An area for cattle to feed in adverse weather.

Corn Cribs: Covered sheds for corn with slatted wooden siding for drying and storing ear corn, (corn on the cob).

Donald G. had a rat terrier dog that caught rats around the corn crib and shocks of grain. Watching a good "rat catching" dog in action was exciting for a boy who had no love for rats.

Hog Boxes: When hogs (swine), were housed away from the barn in pastured areas, there were a number of box-like enclosures perhaps 8 by 10 feet and 5 to 6 feet tall. These small structures were large enough for a sow and her 10 to 12 nursing baby pigs.

Hog House: Greater numbers of maturing hogs were often housed in a separate larger building near the barn.

Hog Trough: A V shaped feeding fixture about 12 inches deep at the center of the V. It was placed on the ground located near the hog house. Liquid and dry feed were poured into these troughs. They were long enough to feed 20 to 30 hogs at a time.

Hog Water Tanks and Hog Feeders: Containers also scattered around in the pasture in sufficient numbers to accommodate a dozen hogs at a time.

Milk House: A place where milk was stored in 10 gallon cans prior to daily shipment.

Silos: The most part were tall round structures used for chopped corn to be fed to cattle (beef and dairy). Chopped corn was harvested when the ears were filled with corn kernels and the stalks were still green. The chopped corn was blown up into the silo by a machine on the ground next to the silo base. Blown corn went up a tall metal pipe about 12 inches in diameter to the top of the silo. The wet corn was allowed to ferment for a few weeks. Silage was fed to the cattle in the fall and winter when there was no grass in the pasture.

Straw Stack: A mound of straw blown from a

threshing machine that separated grain from the straw. The straw was often piled upon a log frame structure with logs on top to form a roof. The open area beneath the straw stack provided temporary shelter for livestock in bad weather.

Tool Shed: A smaller building for storing tools and repairing equipment.

Windmill with Stock Watering Tank: Located in the barn lot. It was filled with water for larger animals, horses and cows and pumped from a well by a windmill or a gasoline engine.

Around the House

House and Home: Many two-story farm houses were built before the ranch type one floor homes popular after World War II. Few were designed with indoor plumbing which was installed later for most farm homes in the late 1930s and early 1940s. Usually there was a kitchen, dining room and bedroom on the first floor, with three or four bedrooms upstairs. The homes were heated with wood/coal stoves downstairs and a wood or kerosene cook stove. The upper bedrooms were heated by the hot air rising through holes about 10 x 12 inches cut in the floor and covered by a register (grill). Often houses built before 1930 had an outside porch or two. Friends usually came to the back porch leading into the kitchen.

Brooder House: A smaller building where baby chicks were fed and confined until they were grown. In the center was a kerosene stove with six heated tubes projecting in a circular pattern above the chicks.

Front and Back Yards: By the 1930s front yards were the formal entrance and were often enclosed by a white picket fence. Back yards continued to be more functional. In the back yard was a clothesline and paths to the garden and outhouse.

Gardens: Most fruit and vegetables eaten by the

farm family came from a garden whose abundance depended upon the tastes of all who labored in them. The labor was provided by women and children, but men also worked in the garden when they had time free from work with the livestock and in the fields.

Chicken House: A building for mature poultry, mostly hens with a few roosters. The chicken house had nests for the hens to lay eggs and a roost for nightly rest. A solid platform to catch the nightly dropping was beneath the frame where the chickens roost. The roosting board droppings were scraped clean with a garden hoe into an old bucket. I can still hear the sound of the scraping hoe mixed with the pungent smell of ammonia in the manure. Chickens were let out to roam during the day.

One Car Garage: Most farm families in the early 30s had only one car. By the end of this decade some farms also had a small truck which was kept in or near the barn.

Smokehouse: This building usually near the house was used for smoke-curing and storing meat.

Springhouse: A building much like the smokehouse was near the house where milk and other perishables were kept by cool water running from a spring that was piped into a trough on the floor or a tank.

Outhouse: A back yard toilet before indoor plumbing. For a school assignment the students were instructed to draw a farm building. Burdette Q. drew an outhouse. A girl seated near his desk, saw his artistic creation. She immediately brought it to the teacher's attention saying "Burdette drew a two holer."

Going to the outhouse in all kinds of weather, though necessary, was not without its challenges. Susanne K. had an additional problem because her mother raised a few turkeys. They were kept behind the house in the outhouse lot. Among the turkeys in the lot was a mean old "tom" turkey. He would chase Susanne and flop her legs with his wings. Susanne stopped all this nonsense with a broom handle kept by the gate for her protection. Old "tom" turkey would not bother her when threatened by the broom handle.

Wood/Coal Shed: This building was attached to or near the house. Winter fuel was stored here as well as lawn and garden tools and other tools used around the house. In some places the building also housed a water pump and was used for washing clothes.

Woods, Orchards and a Stream: Woods provided lumber for construction, fencing and fuel. Orchards often had a variety of fruit trees probably planted generations earlier but often neglected after the 1930s because of the time and expense of upkeep. Those who took care of larger orchards did so for the sale of the fruit. Streams were basic to watering livestock. All of the above proved to be wonderful areas of recreation. Adults as well as children cherished times of climbing a tree, looking for a good apple and wading in a meandering stream.

That's The Way It Is
Family Farming—a way of life

Those interviewed agreed the family farm is much more than a place to make a living; it is a way of life where all the members of the family have a vital part to play in the operation of the family farm enterprise. It was a great place to raise children.

With the eyes of an enlightened heart we see beyond the barn and beyond the faded memories of the farmstead. We see the heart, mind and soul of farm family members and their neighbors. A rural community was more than people in geographic proximity. The divining essence of a rural community was spelled out in the relationships where competition and cooperation were in wholesome balance for the good of all.

The family farm is a specific area of real estate where family members work together to make a living and embrace life. Their primary labor is to till, plant, cultivate and harvest food and fiber for personal use and market for financial gain. Their labors usually include care and feeding of livestock and poultry, also for family sustenance and market produce. Farm families prize the challenge of freedom to be personally responsible for their labor management. The fruit of their labors is enriched by an intangible bonding of members of the family, as well as with other farm families, the livestock and the land. The satisfaction of work on the family farm is enhanced by pride of ownership and becomes entwined with the pleasure of recreation, reverence for the world of nature and appreciation of self-worth. For many, working with neighbors, land and livestock is also a bonding with God.

As stated in the introduction, the people interviewed for this book were not selected at random. Those who accepted the invitation to share their memories of a diversified family farm did so because they treasured their stories and were eager to hear others. The sharing was usually done with two to six others. Not unlike hunting morel mushrooms, one story popped into memory only to be followed by another and another. A grand time was had by all.

When I grow up

"What are you going to do when you grow up?" is one of the unfair questions often asked of children. What's wrong with enjoying being a child? The hidden supposition is that of the degradation of childhood. "Aren't you ever going to grow up?" is the cry of many a frustrated parent. At some point in their growth, children venture from the satisfactions of the present to pondering the future. In the turbulence of adolescence many children share with their peers dreams of a time when they are "grown up." Being independent and your own boss was the dream but at times the down fall of living on a family farm.

I'll never marry a farmer

Somewhere in the "teeter-totter" years between the upper grades and early teens, girls watch boys and boys look back. At school on a day when it was warm enough to be outside Dotty R. and her twin friends Helen and Hazel were walking and watching. At such a time the hidden secrets of the heart rise like cream to the top of fresh milk. Though they would have denied it if accused, they were talking about boys. The twins lived on a family farm and had eyes for the farm boys. Dotty lived near town and said she would never marry a farmer. She planned to live in the city. Helen and Hazel were rooted in the country. Both hoped to marry a farmer. As often happens with early plans, the twins married men who worked and lived in the city. When they visited Dotty, they saw what she had learned to like about being a farmer's wife.

A chicken parade

If you have never seen a "chicken parade," Pat H. can tell you all about it. All one needs is an imagination, chickens and an ear of corn. Since there were no chickens where she lived in town, it did not happen there. Pat had grandparents who lived in the country. That's where she discovered the art of parading chickens. This was one of the fun things that foster fond memories of the many times Pat stayed on the farm with her Jones grandparents. Gathering eggs was one of the chores given to her. At times she took an ear of corn with her to shell for the chickens to eat. When shelling the corn, the grains fell one grain at a time at her feet. Pat was not afraid of the chickens gathered at her feet, but she noticed when she backed away still shelling the corn the chickens followed the grains. It was then that it became a game. As she walked around the chicken lot shelling the corn the chickens followed in a line gobbling up grain after grain. The string of chickens following her picking up each kernel of corn formed her "Chicken Parade."

Slopping the hogs with grandpa was no big deal for her. On the platform above the bull pen, she could reach over and pet the bull. She showed off by doing this in the presence of some of her city friends. They were impressed and thought she was so brave. She grew up in town, but her heart was in the country. Pat wanted to marry a farmer, and she did. She dearly loved David, a fine young man she met at Quaker Knoll church camp. Could it be the fact that he was a farmer helped Pat accept his proposal of marriage?

No better buddy

Lois H. had a "buddy," the best kind of one. Grandmother Maggie was her "buddy." Lois grew up in east central Kentucky. After WWII her parents lived with her mother's parents. Lois was close to her grandmother. Grandmother Maggie took Lois with her to many visits with older relatives and friends in the community. Lois considers herself fortunate to have been raised in a community of family and neighbors. Her father worked away from the farm. It was not a large farm, but they were well fed with the farm's produce. With a garden, livestock to butcher for meat, chickens for meat and eggs and cows for milk, most of their meal needs came from the land and livestock. Families in the community took responsibility for each other. Butchering hogs with the help of neighbors was a given. Men and women as well as children came together to labor and visit. The school and church were her focal points of socialization and of vital importance in the whole community.

    Meal time was a significant time of family worship and fellowship. Breakfast, in some ways, was the most important meal of the day. Lois found this to be true once when she failed to appear at the appointed time. When Grandmother Maggie came to check on her, Lois decided she was sick. Being sick seemed to be better justification for missing breakfast than forgetting or dawdling. So Lois claimed she was sick. If you are sick in grandma's house, sick enough to miss breakfast, you spent the day in bed. If you were sick, you could get well faster if you did not read a book, listen to the radio or watch the black and white TV. Lois decided she was not that sick. Grandma Maggie felt otherwise. Lois made it to breakfast from then on. Grandma was still her "buddy." Now a grandmother herself, Lois has nothing but praise for the small family farm and her beloved "buddy."

A real "hoosier" cowboy

    At age five Press A. became a cow-milker. His dad Herbert, was a toolmaker in Connersville, Indiana, but was born on a farm. With fond memories of the farm, Herbert rented a farm and moved his family to the country. He believed a farm would be a better place to raise his six children. The move to the country on a farm meant every family member had chores to do. With the challenge of being the youngest and the confidence of a five year old, Press determined to find his place of labor. Why couldn't he milk a cow? His brothers did. They thought he was too small, but he knew he could do it. Though it did not come easy for him at first, he finally learned the art of cow milking.

It is easy to tell whether or not a person is an experienced cow-milker. The first clue begins shortly after he or she sits down on a low stool with a galvanized metal bucket under the udder of the cow. With experienced cow-milkers, the rhythmic ring of alternating streams of milk striking the bottom of the galvanized metal bucket soon begins to diminish as the accumulation of milk covers the metal and continues to fill the pail. The speed of the subsiding sound of milk on metal depends upon the strength of the hands and the experience of the person doing the milking. First time milkers usually had little success in getting much of a stream of milk as they squeezed the teat of a cow with each hand.

Even young eager cow-milkers realized it's not as easy as first appears. Cows express their displeasure in any number of annoying ways such as putting their foot in the bucket of milk, or using their manure encrusted "fly swatter" tail as a club to the head of the cow-milker. The last insult, and most annoying, hits you when a cow kicks the bucket over. The spilt milk is truly worth crying over. However, Press soon discovered he had the pull for the job.

Once a competent cow-milker and mature five years old, Press decided he needed to make a change. A very simple plan came to his mind. At the supper table with the audience of his whole family, he gave notice that he'd had it with the cows, and was quitting milking when he became six. As an adult when he had the freedom to make a vocational choice, what did he do? You guessed it. Press farmed and milked cows.

A great place to raise children

No mention is made of children in the "Garden of Eden" for good reason; there were no chores for children. Hard work did not start until Adam and Eve were evicted because they violated their lease. By far most of the people interviewed for this book have high praise for the challenge and rewards of the chores they had to do on the family farm. Most admit it was not always fun time, but those testing times left to memory have faded into times of satisfaction. Children ordinarily in both town and country were expected to share in family labor. On the farm parents saw more of their children since they worked with them and had a greater measure of supervision over their maturing.

David H. remembers the conversations they had at meal time when the family was all together. At such times the family sharing often related to some phase of the operation of the family farm. This was a regular everyday occurrence, not just on special days or occasions.

Something of special interest occurred nearly every day. A newborn calf, pig or lamb, the pigs found a hole in the fence, the crops looked good but they needed rain, someone forgot to get the eggs, etc.

Children on a family farm with a diversity of crops and livestock had a greater variety of chores to be done. The choice of chores was not always left up to the child, but there was something of interest for every age. Farm family children learned to take responsibility in keeping with their interests and ability. Every family member had a significant place in the success and happiness of the whole family. Each in their own way was a farmer.

For Dick G. and others, farm chores whether in the house, garden, barn or field, were undertaken with a sense of personal responsibility and commitment to his family and neighbors. He, his brother Don, and his father Clarence, took turns doing the chores. Every third week it was Dick's turn to feed and care for two dozen sheep and about the same number of beef cattle. When his mother Margaret called and said, "It's time to get up," Dick knew it was his turn to care for the livestock. On a cold and windy winter morning, he wished it was not his turn, but mother's call and Dick's commitment had prior claim over the soft warm bed.

Life spiced with variety

Children who lived on diversified family farms were exposed to a wide variety of life experiences. Most were afforded an in-depth understanding and appreciation for the cycle of life, from animal mating, to birth, growth and the final slaughter of the animal or fowl for food or sale. Children as well as adults were attracted to the adorable antics of the young that grew to a life-long affection for their animals. Family farm children enjoyed their own year-long petting zoo. However, they also knew the reality of killing livestock for food and by-products plus the pain of losing a pet by death or market.

Many adults have a vivid memory of the fountain of feelings that erupted when taught to kill a chicken. Few things measure up to the twisting of innocence like seeing a headless hen or rooster flopping out of control far longer than one might expect.

Dorothy D.'s grandson could not believe she actually killed a chicken. He found it hard to think of her hanging a chicken by the head on a clothes line and cutting its head off with a knife. Nor could he picture his grandmother stepping back to keep from being covered by the blood spattered by the dying, flopping, headless chicken. Much like fresh tree ripened fruit, chicken killed, de-feathered, cut up, fried and eaten the same day had much better flavor than chicken frozen or kept days in a refrigerator.

Today people who reside in the country have much the same lifestyle as those who live in the towns and cities. Children of this day live in a world vastly different from the world of their grandparents. Only a few children today have any experience of caring for livestock or producing and processing food for the family. Families of the farm now trade at the same supermarkets and stores as do the families in urban areas. Like many families who are busy with things to do and places to go, having all the family together for every meal, is the exception and not the rule.

Dorothy D. concluded life was simpler in the 30s and before. They had fewer options back then. Most of the food for the family and the livestock was grown and processed on the farm. Basic food was eaten fresh in season and canned or dried out of season. Man, horse and vegetation provided most of the energy necessary for the functioning of the farm. With few exceptions people were friendly, trusting and dependent upon each other for service and socialization. It was a good way to live and raise children. Today we have become more separated because of having so many laborsaving devices that replace a person. The absence of that helping neighbor or friend has had a lot to do with the breakup of community. Dorothy's motto was and still is, "Love many and trust few, but always paddle your own canoe." Our challenge is to reclaim the balance of private and public in the setting of community.

Confining routine—it pays the bills—no regrets

Ray C. liked to milk. He grew up on a dairy farm with the security of the routine. Ray knew no matter what came up the cows had to be milked twice a day, 7 days a week, 365 days a year. Income from the crops came once a year. With dairying, Ray and his wife June could count on receiving a milk check twice a month year around. After a problem week when things were not going well, Ray promised June, "One of these days, we are going to take off a week for a vacation." It did not happen the next week or next month but Ray kept his promise. When Ray and June had their own farming operation and had been married 13 years, they decided this was the time to take that week off from milking and go for a vacation.

Ray and his father milked by hand until electric

power was brought to their area. Ray was 8 years old when they got a milking machine. His mother milked when Ray went to college. June helped Ray with the milking after they were married. For a while she got up with him at 4:00 a.m. and went out to get the cows while Ray was getting ready for the morning milking. After a time of this stumbling around in the dark, they decided it was not necessary to get up quite that early just because his dad did.

Ray's Dad Hower was dedicated to work. He had no hobbies. Working in the fields and with the livestock was his recreation and greatest source of satisfaction. When Ray mentioned their plans to get away for a week, his dad saw no merit or need of a vacation.

Hower's comment was, "I've always found that if a man is going to be successful, he has to stay right at it." Never-the-less, Ray and June took the vacation they deserved. It is rare for successful dairy farmers, but "never-the-less" has its day.

Farming has been an enjoyable and rewarding business and "Way of Life" for Ray and June and their family. In Ray's words, "I don't regret it one bit. But we have witnessed the gradual, or maybe rapid, loss of the life style part to see it become 'Big Business'. We have been subtly prodded to 'get bigger or get out'. A few have accepted that challenge, but more have 'gotten out' or have taken outside employment to supplement farm income which generally leads to full-time off farm employment and farming becomes a hobby.

The goal of the family farmer, it seems to me, should be year around full-time employment for himself and the family labor. This is where livestock fitted into the program in the 1930s through the 1960s. As livestock was dropped from the operation because of low return, high cost of fencing and confinement buildings, crop rotations disappeared too. What were we left with? Corn and soybeans and more corn and soybeans, and that type of operation seem to require lots of acres and big machinery. The farmer may go night and day for maybe a month in the spring, (planting) and a month in the fall (harvesting). Without livestock we are back to less than full-time employment for maybe 9-10 months of the year. That is far from the year around full-time employment for even the smaller diversified family farms in the 1930s."

The green grass of home

Burdette Q. loved to live and work on the farm. It always lifted his spirits to see young livestock freely romping in green pastures. Even animals well up in years kicked up their heels when turned loose from a stall or confining pen. For them and Burdette, life on the farm was good. Anyone who visits him and brings up the subject will be treated to a pride-filled picture presentation of grandchildren and their prize winning animals at the fairs.

Bobby Q. has a hobby collecting old farm machinery. This adult recreation brings back fond memories of a time when each of these collectables contributed to the "good life" on their family farm.

Soul soil for all seasons

Margaret H. enjoyed the crops more than the livestock. She had no special dislike for livestock on the farm if they behaved and kept to their appointed space. It is not always easy to get livestock to agree to these terms. Once the seeds were planted in the ground, she could sleep at night with the knowledge they would still be there when she woke up. Not so with livestock that may or may not stay overnight in their resting place. Her place and time of great pleasure was grounded in the soil heralded by the seasons. Nothing was dull or uninteresting about the seasonal routine from preparation for planting, cultivating, harvesting and working the soil for another season. Each crop and season served to support her gratitude to God for their home and farm.

Homegrown—reason not to buy

Their mother Mary H. raised most of the food they ate year around in the garden. She canned garden produce for the winter months. With livestock on the farm they had an ample supply of meat and milk. Mary purchased very little food. The small amount she bought was at stores where she could get the most for her money. This degree of subsistence living is possible today on a small family farm. Jim, David and Becky H. know it to be true.

When it comes to money there is a wide deep gulf between "not enough" and "very little." It is hard to imagine living with very little money with near total dependence upon money to purchase goods and services. This chasm grows in width and depth when borrowing off the future is seen as the only way to meet a present crisis of "not enough." Prosperous people struggling in the land of "not quite enough" have lost sight of those on the other side of the chasm making it on "very little."

Making the most of very little

Born and raised in Breathitt County, Kentucky, Gene T. knows what we have given up in our climb to be financially successful. They had very little money, but they didn't need much. Life's essentials, food, clothing and shelter were the produce of their ingenuity and cooperative labor with livestock and the land. Most of their food came from the garden or livestock and poultry. For cooking and heating, trees were cut and coal was dug from veins in the hills which rose up from the valley floor. Neighbors worked together for the satisfaction of the relationship as well as completing a task that one person could not do very well alone. The line drawn between family and neighbors was erased by frequent encounters and the demands of survival. Community existed extending a life-long helping hand from the birthing of a child, to the making of a casket and digging the grave.

The "very little" money came from selling produce of the land and livestock for certain food and clothing items as the need arose. Kerosene was needed for lamps. Since there were few roads, they had limited investment in cars and gasoline. Their main source of energy for transportation was the horse. Horses were used to plow and cultivate the garden. Since there was not a lot of flat ground they did not have much grain farming. People in that area raised only what they and their animals needed to live. It was a good life.

Everybody had porches and a porch swing. It was a place to get out of a hot house and watch the world outside their door. Care of the garden began with the first morning light. Work was done in the cool of the morning. Then it was time to sit in the swing on the porch and relax. And it was time to tell tales and jokes like these he shared with me. I enjoyed them. You will have to ask Gene if they are true.

Can you take a joke?

A guy was driving down south and stopped to get gasoline at a country store. He noticed a big watermelon on top of a pop cooler. It must have weighed 20 pounds. The visitor also notices a boy standing by the cooler and thought he'd have some fun. He told the boy he would give him $5 if he could eat that whole watermelon. The boy thought he could but said "give me about 20 minutes to run home first." The man was not in a hurry so he thought it would be worth the wait. True to his word the boy came back as he said he would. He got a knife, cut the melon open and ate every bite of it. The man said he'd keep his promise and gave the boy the $5 but he first asked the kid what he wanted to go home for. The boy told him, "my dad had a watermelon at home about the same size and I ate it 'cause I know'd if I could eat dad's, I could eat this one."

Confronted by a sailing sack of coal

Gene's uncle and others got coal out of the side of the hill. They usually let it slide down the hill and picked it up where it stopped. His uncle rigged up a wire slide from a tree at the top near the mouth of the mine to a tree down at the bottom where they needed the coal. He took big bags that coffee came in, filled them with coal, hung them on a pulley on the wire and let the bag of coal roll down the wire. Once in late evening a man came walking by. He couldn't see the wire but he got quite a scare when he saw the sack of coal that seemed to be sailing in the air down toward him.

It may not sound much like the family farm some of us knew in this area in the 1930s, and certainly not today, but Gene knows it was the best of living without much money.

Surprise we have to move

"Why do we have to leave our house? Why is daddy not happy, is he sick? Why is grandmother not staying with us?" Why? – Why? – Why? How do you explain family problems to a five year old? Dorothy M. did not remember her grandfather Walter who died before her parents moved to the farm. When he died her grandmother Blanche asked Dorothy's father Isaac to come home and take over the operation of the farm at the top of the hill west of Wilmington, Ohio above Todd's Fork. Isaac lived and worked in Cincinnati, Ohio, but agreed to make the move with his family back to Clinton County. He took over the farm expecting it to be a long-time arrangement. Dorothy was five when her grandmother remarried, moved away and sold the farm. The surprise move came during the depression and was a devastating blow to Isaac and his family. The times were turbulent and testing. Dorothy's father had to go in search of a farm to rent to make a living. The next spring he had the option to renew the rental of the farm, or look for another one. Isaac elected to find a better farm. Finding another farm brought all the anxiety of adjustment and moving again. Every spring for a few years, Isaac hunted for a better farm to rent. Painful as it was, each year proved a forward step up the ladder to ownership of a farm. Moving each year was quite a challenge for Dorothy's mother Helen and Dorothy as well. Now as she looks back on these days of challenge and change, Dorothy still acknowledges her appreciation for farming as a way of life.

What is this living on the farm bit?

Dorothy D.'s memory of her very early years on a

farm remains hidden in the cloudy predawn days of the very young. In the turbulent times of the depression, her father did not have enough money to keep their farm. Dorothy's father Frank and mother Mary with two small children had to look for another place to work and live. A few months later tragedy struck another blow when Dorothy's mother died of pneumonia. In desperation and burdened with grief Frank knew he needed help to care for his two young children. He turned to his sister Inez and her husband Earl for help. Dorothy was taken to Cincinnati to live with aunt, uncle and cousins. Her older sister went to live with her Grandmother Ida in Cherry Fork, Ohio.

Life during the early 1930s was quite different for those who lived in the cities than on a farm. Homes in the city had electricity, running water and food from nearby grocery stores. Dorothy's aunt and uncle were in far better condition to care for Dorothy than her father without a wife or job. In the diary of her memory, at least Dorothy lived comfortably in Cincinnati with her aunt and uncle and their two children.

All went well until she was eight. Little did she know or understand what was ahead for her. Her aunt had two children and was expecting another child, when she learned Dorothy's father Frank had remarried. Frank received a letter from his sister that said, "Remember, you promised to come and get Dorothy if you remarried. We need you to come now and get her."

In keeping with his promise, Dorothy's father brought her to the farm where he and Dorothy's stepmother lived. This was a very traumatic time for Dorothy. She now had to live with two people whom she did not know, in a place where she had not been, and on a farm. Life on the farm was so foreign to her she hated it. Animals were around everywhere. The food they ate on the farm did not come from the grocery store in boxes or cans. How could food be clean when it came from the dirt of the garden? Eggs came from chickens and milk from a cow instead of a bottle in the ice-box! This city girl had a lot to learn about life on a farm.

Dorothy spent many unhappy weeks of adjustment, but given time she came to accept and enjoy life on the farm. It's hard not to like little lambs, calves, pigs and chickens. The wonder of seeds in warm moist earth growing to be splendid flowers or tasty vegetables helped Dorothy find happiness away from the luxuries of city living. Her times of tears soaked into the comfort of loving parents and the good earth.

As an adult living away from home, she got homesick for the farm. Whenever she could, Dorothy came back to the farm. One of these times when she came home to renew her spirit, her dad with a teasing laugh asked, "Are you here to plant cornfield beans?" Dorothy said, "Yes I am!" She took off her shoes and danced barefoot out in the cornfield at one with the soil that enriched her soul.

Even though I've had it

For those who were born in the mid 1950s and later, there is probably no better way to understand the ups and downs of "making it" in the 1930s than listening to those who lived during challenging times. Mike D. had a unique opportunity learning of those times of trouble in a one-to-one conversation with his grandfather. Grandpa Frank accompanied him when Mike was a "go-for," (the designated driver), to go for supplies for the family hardware store. Designated drivers get along better if they have company to help fill the time it takes to go to a distant city for hardware supplies. All Mike had to do was ask his grandpa about the times when he was growing up. The miles flew by as grandpa told almost endless stories of his life journey. The way his grandpa met some weighted tragedies is an inspiration for Mike and many who have heard the story.

The list of challenges seemed far too great for one person to bear. Grandpa Frank lost the farm in the 1930s. Six weeks later his wife caught pneumonia and died. His two daughters, Mike's mother and her sister had to go and live with others because Mike's grandpa had to go to Cincinnati to look for work. He was living on the street and eating in a soup kitchen because he could not find work. Just six weeks later he learned that his oldest daughter died of burns she received when she fell into a hot kettle of lard to be used for making soap. Mike was stunned by the magnitude of his tragedies. He marveled at his grandfather's friendly manner and lack of bitterness and asked him how he was able to deal with all this sorrow and suffering. The grandfather's answer made a lasting impression on the grandson. "Son, some days you wake up and life is hell. Other days you wake up and life is good. The secret is, you just keep waking up." Grandpa Frank lived to be 91.

Country to town and country again

No running water and limited heat in a cold bedroom has its down side even for a preschooler. Eleanor G. at age five had forgotten much of the dark side of living in such a farm house soon after moving to the city of Chillicothe, Ohio. Eleanor's father was a school teacher. Her parents had their roots in the country. The first five years of her life she lived in a farm house without electricity. Limitations of comfort and convenience prompted a move to Chillicothe where they had electricity. Here Eleanor enjoyed the benefits that went with it. No more carrying dim lighted lamps from room to room, pumping water by hand, waking to cold bedrooms, shivering while trying to dress and no more trips to the outhouse. City sidewalks were meant for roller-skating and riding her bicycle. With the bicycle Eleanor could go uptown to see a movie with friends and get ice cream or candy. She was happy attending Mount Logan School. What more could she want?

When her father Ray announced the family would be moving once again out of town to another farm, it was far from good news for Eleanor. At the age of twelve, the previous life on the farm held few pleasant memories for her.

What teenage girl wants to be carted off from the security and comfort of a place with her peers? Her father wanted to make the move because he thought city life was having a bad influence on his five children. Under protest, Eleanor the middle child, between two brothers two sisters, moved back to the country. Once again she was on a farm without electricity.

It was a labor-intensive small farm. Her father Ray taught nights at the Chillicothe Corrections Institution. His being employed off the farm meant chores for the children. Both her sisters preferred to work in the house with their mother Elsie. This was a plus for Eleanor because her love of being outdoors and working with her father. She remembers riding and operating the wheat binder when she was 14. On the farm they had two acres of tomatoes and blackberry patches. The children were involved in the harvest. Blackberries were sold in Greenfield, Ohio. Tomatoes were sold in Washington Court House, Ohio. Eleanor had the job of driving the truck taking the farm produce into market.

Even though it was a small farm, it was diversified. On their farm they had cows Eleanor helped milk by hand. She liked the horses. When it came time to milk, Eleanor rode a horse to round up the cows. Being high upon a horse in command of both the horse and carefree cows was exhilarating. Eleanor also rode the horse pulling the one-shear plow while her father plowed the garden. Besides the garden, they had a grape arbor. Her mother and the girls made most of their own juice and jelly, as well as most of their clothes. It wasn't exactly Chillicothe, but on the farm Eleanor discovered deep satisfactions that would serve her well in her adult life. Like her father, mother, uncle and aunt, Eleanor became a teacher. Like her father, she was a farmer. Eleanor married a farmer and raised children on a family farm.

Count your blessings

Darell F. grew up on a diversified family farm. He enjoyed farming immensely and wishes more people, especially today, could grow up on such a farm. Darell credits his family unity and satisfaction of working with the land and livestock as gifts from God the creator.

When the cock goes off

The old saying "Early to bed, early to rise, makes a man healthy, wealthy and wise," was not as much a guiding goal for Donald D.'s father Loren as it was the fruits born to those who loved the labor of a family farm. Donald's dad got up early even before the predawn wakeup call of the rooster. If he chanced to see a neighbor's light when he got up he'd get up earlier the next morning. What made him do it? Making money was not the main motivation. He prized his freedom to be his own manager and loved to work with livestock and the land. Donald's father took pride in being a hard working farmer. To say he enjoyed working on the farm is a bit of an understatement. Feeding and caring for animals took time morning and night. Care of livestock plus the seasonal labor of planting, cultivating and harvesting the produce of the fields meant long days and short nights. Long hours and hard labor had no dread for Loren. However, after the animals were cared for on Sunday he went to church. Sunday was his day of rest and fellowship with others in the neighborhood church. The satisfaction of Donald's family was rooted in a firm faith in the goodness of God evidenced in all creation and the loving support of neighbors and friends.

Rosalee W. and others noted many rural people went to church on Sunday. Very few places of business were open on Sunday. Instead of eating in a restaurant on Sunday, farm families often had picnics with neighbors and relatives. In the colder months, the families still got together for Sunday dinner with their neighbors. At times families came together for dinner and an afternoon of visiting after worship. Planning in advance was not necessary. Many unexpected visits were the result of a simple invitation of, "Come eat with us." Children often initiated the invitation with parental consent.

Hold Your Horses
Farming With Horses and Equipment

*The use of reapers contributed to a shift from oxen to horses for motive power.
Horses were more nimble than oxen, were faster and moved at a steadier pace,
all factors that made them especially appropriate for machinery.*
David Danbom—*Born in the Country*

It seems a bit ironic that we rate the power of an engine by Horse Power. Today we are more interested in horse "pleasure" than horse "power." It is good to know people with great stories of both the power and pleasure of horses. While it may be impractical for most people to work with horses today, we all pay dearly for engine horse power. True horse power came from the energy of oats, grass and hay 80 years ago. The price was hard to beat.

Old Vic was blind

When a person is blind, he or she may be blessed by the guidance of a dog trained to be eyes for the blind. That was not to be for old Vic. He had no money or use for a guide dog. Self-pity had no place for him. Those who knew him best claimed he was always gentle, steady and calm. Old Vic did not have the help of a support group for the visually challenged, because Vic was a horse. One might assume life had little to offer an old blind horse. A compassionate owner might continue to give such an animal shelter and nourishment for the rest of its days. Others would justify disposal claiming they were, "putting him out of his misery." Neither of these two options was in place for old Vic. Vic was a valued and productive member of a family farm.

It was a spring day perfect for plowing. The air was cool and brisk, but the sun climbing up above the eastern tree line sealed the promise of a warm and glorious day. It was time for turning the emerging green growth over to expose the rich dark soil. Donald D. felt the warm inner glow of anticipation of a day of satisfaction working with the three horses. Dewey, Nell and Vic were harnessed to pull the sulky plow. Dewey was on the right, Nell in the middle and Vic on the left in this three horse hitch arrangement. While some plows were walk behind, the sulky had a seat for the driver. Donald held the reins, but the horses knew the way. Blind Vic never missed a step on the farm lane to the back forty, (forty acres of ground to be plowed before further preparation for seeding). The muted thud of twelve hooves was accompanied by the mechanical sound of turning plow wheels with an occasional snap of a twig or bang of a rock that had worked its way up to the surface. At first, all three horses had equal footing on the unplowed ground. When there was a furrow to follow, Dewey and Nell walked on the unplowed ground. Old Vic walked in the furrow between the plowed and the unplowed ground.

Much to the delight of a continuous following of birds of the field, rich black earth was turned up and over by the 14 inch plow shear. They were noisy in their gratitude to Donald and the horses for a grand banquet of bugs, grubs and worms exposed from their hidden home in the freshly turned soil.

For Donald this was living in the finest sense of the word. It was a symphony of the good life. To be so in tune with nature lifted his spirit to praise God and all creation. The noise of machinery with the attention required for its oversight while in operation would surely drown out the satisfactions of working with horses. Working with machinery he would not hear the snap of roots severed by the plow shear or the swish of the soil as its productive side was turned to the light of day. Nor would he hear excited chatter of the birds in search of food, or the whisper of a cooling breeze passing through the trees in the fence row. There would be no bonding rest with a tractor as with the team at the end of a long strenuous stretch across the field. Lastly there is no gratitude expressed by inanimate machinery comparable to that of a horse when brushed and fed at day's end. Another day crowned with a spectacular sunset, was a fitting celebration of the union of a weary body and satisfied soul.

When a horse or two or a mule will do

The roads were mud in the winter and spring in the mountains of Breathitt County, Kentucky. At times people got their car or truck stuck in the mud of the road near Gene's house. When they came up to the door and asked for help, Gene's dad Jack and the horse were ready to go. Jack kept a work horse harnessed even at night. Some paid him, some didn't. Families had one or two horses and some had a mule. Often families had two kinds of horses, one for pleasure and one for work.

We are truly blessed when living and labor falls into an orbit of comfortable routine. Without the ritual of the routine on a family farm, not much would be accomplished. However, life can be drab and dull with no detracting detours. Dull routine is not a problem on the farm. Horses break down fences and gates, cows get into the corn, it rains when you have hay drying in the field and stove pipes from the cook stove are known to separate at a joining seam leaving a covering of soot all over the house. Sometimes when routine is called to a halt, the result is a story that's too good not to tell. Gene's dad had such an experience when he went out to check on a prized mare about to give birth.

A mare named Bert was about to foal. Jack worked on the railroad and always carried a prized pocket watch in a chest pocket of his coveralls. He would take out his pocket watch from time to time and go out to check on the mare. This time when he went out to check, the ground was a bit frozen so he had to give the barn door a jerk. Jack discovered the foal had arrived in a way totally unexpected and unacceptable to him. The newborn now standing on wobbly feet was frightened by the sudden noise of the door. With a jump and kick the foal planted a foot squarely on the chest of Gene's dad. While it did not knock Jack off his feet, it did hurt. The greater pain was not in the chest. The foal was able to make time stand still by striking and breaking his prized pocket watch. That's what really hurt. When his dad came back into the house it was clear he was not too happy about the untimely foalish action. (pun mine).

A bull-headed horse

Fred is more shy and timid. Ed's the outgoing one, but Ed can get "bull headed." Jeff H. is not talking about twin boys in the neighborhood. Fred and Ed are the names of a pair of mature Percheron draft horses. In a way they are "hobby-horses" since working with the horses is one of Jeff's hobbies, but Fred and Ed earn their keep. They do a significant amount of labor on Jeff's farm at times when horse power is more relevant to the task than tractor power.

On the hobby side, he wanted to slow down a bit and go back to the way they used to work with horses on the family farm before the takeover of tractors. Jeff enjoys working horses savoring the satisfactions and challenges wiped out by tractor racket. Working with horses without the noise of the tractor one could hear a lot of other peaceful sounds of nature coming from birds, animals, insects and the wind. Plowing with horses combines the swish of the soil sliding over the plow mold board with the thud of the horse's feet.

From time to time the pop of shearing a large root will mingle with the noise of the horse harnesses.

Jeff warns us not to think horses are docile creatures subject to man's every whim. It doesn't pay to get angry with his horse Ed. The horse is a lot bigger. When Ed the horse gets destructive and you get worked up, if you settle down, he will settle down. Getting them to do what you want them to do is a different kind of challenge. It's a mind game. A mindless machine will run as long as it has fuel and is in working condition. A horse can go full steam, but if he changes his mind and doesn't want to do it, he will quit. You have to figure out how to get them to see it your way. Jeff likes it that way. Sounds like good advice for raising children. If children get the same message there is cause for rejoicing.

Nothing will get your goat like a smart horse

In 1939 Gene T.'s Dad Jack made trips from Kentucky with rough lumber to have it planed at a mill in Fayetteville, Ohio. He sold the planed lumber to any who had a need for it. While here in Ohio, he had an opportunity to buy a farm near Fayetteville. To help in the farm he bought a team of horses that were not well trained, but they did the work for him. One of the horses was smarter and lazier than the other. When plowing this horse would keep the traces just a bit tight so it looked like he was working, but when the other horse stopped, everything stopped. Is there a lesson on working with others here?

When a guy really needs the help of a horse

Mike D.'s grandfather had a cousin known as Squire who lived with different family members from time to time. One winter Squire came to live with Grandfather Young and family. When he came home with a horse and buckboard wagon, no one knew what use he had with such an outfit. Squire would occasionally take off with the horse and wagon. The family did not know where he went. He was inclined to drink a little too much alcohol. They figured it was a "none of your business" mystery and let it go at that.

This continued well into the summer. One morning Squire was found to be asleep in the wagon, with the horse patiently waiting the schedule of activities for the day. When it all came to light, the responsible one was the horse. Squire had driven to Seaman, Ohio to an establishment of alcoholic refreshment. When he was well into a time of relaxation and repose, he instructed his more sober drinking buddies saying, "Put me in the back of the wagon and slap the horse." Long before man saw it as a prudent arrangement, the horse had enough sense to ignore the unmanned reins and be the "designated driver."

Try these shoes for size—both pair

Many farmers did their own horse shoeing and doctoring. The art of caring for horses was passed down from generation to generation. They took good care of their horses, fed them and wiped them down when they had worked up a sweat. Both of Gene T.'s grandfathers had blacksmithing skills. They looked after the horses' feet and made their own horse shoes.

The Shetland pony they bought from Frisch's Big Boy Horse Farm needed shoes. Gene got the smallest shoes he could find. Gene's father Jack watched as Gene put them on her front feet. He bent the nails so they would turn out rather than go in and dig into the flesh. The pony didn't seem to be bothered when Gene put on the front shoes. But when he started to do one of the hind feet, it was a different matter. He picked up the foot and started to work on it. The foot began getting heavier and harder to hold. Gene's dad Jack was grinning. He knew what was happening. He told Gene, "She's shifting her weight on you. She'll have you on the ground if you don't watch her." When they finished they took the pony out to the hard surface road and she stopped dead in her shoes. The noise of the shoes clomping on the hard blacktop was foreign to her. The pony's work was pulling a wagon with children in it. She finally accepted her new shoes and pulled the wagon.

On having a "less than perfect day"

Donald D. is very thankful all farming days were not as challenging as the day he set out to harrow one of their plowed fields. The field was on the other side of the creek, separated from the barn and farmstead. For those who've never heard of a harrow, it is a flat, about five foot square, implement with rows of 6 inch spikes for breaking up clods of dirt in plowed ground. It is a giant rake, something like a large slotted storm drain cover with teeth sticking underneath. This day two harrows were hooked together side by side. A heavy 10 foot pole was chained to the back of the harrows. The pole, or "drag" as it was called, helped break up the dirt clods and smooth out the loose soil to make a better seed bed. To get to the field across the creek with the harrow pulled by three horses, Donald had to cross over on a bridge. His father built the bridge ten feet or more above the stream with sturdy oak planks and supporting timbers. Time-after-time over the years, horses, men and machinery crossed safely over this time-saving portal to the land beyond the stream. Today's passage was definitely different. The trusted team of three, Dewey, Nell and old Vic, were harnessed for the task. As usual Nell was flanked by Dewey on the right and Vic on the left when they were hitched to the harrow and drag. When transported from field to field, the harrow spikes were turned back parallel to the frame so they would not dig up the road. Donald was looking forward to another satisfying day working with the horses. They too seemed eager to be in harness and out of the barn lot as they began the ¾ mile walk to the bridge over the stream and on to the work in the field beyond.

It had been a good morning. The harrowed field

now rested ready for the sowing of seed in anticipation of another bountiful crop. Time passed quickly as Don and the horses began the tromp, tromp, tromp back over the bridge. To Don's knowledge the bridge housed no mischievous trolls of folklore. However, given what happened next, might give one cause for wonder. Some part of the harrow, perhaps one of the spikes, caught momentarily on part of the bridge. The unexpected jerk of the catch threw Don and the horses a bit off balance. It all happened in a fraction of a second. Dewey and Nell were tossed against Vic. The next thing Donald saw was blind old Vic separated from the other two. The poor horse had fallen off the bridge. "Oh no, no, not old Vic!" raced through his mind as he gazed in wonder at the faithful horse in the streambed 10 feet below. Nothing in his experience with horses compared to what happened to old Vic. Donald had expected to see the horse thrashing about in an attempt to climb up the stream-bank, or even worse, a broken, twisted heap in the streambed. Years later it is still hard to believe what he saw. Vic, an old blind horse with harness twisted and torn, was standing in the stream as if nothing had happened. He seemed to be waiting patiently for the next tug of the reins and command to move up and out. We will never know if it was divine help for horses, or the innate triumphant spirit of a blind horse that had learned to cope with the unexpected. Perhaps it was a bit of both. Vic was apparently unhurt. The horse had landed feet first in the soft sandy mud of the streambed. Donald hurried to the aid of his old friend. In spite of tremendous effort, he was unable to lead old Vic up the slippery slope of the stream-bank. With haste he tied the other horses to the bridge and rushed the ¾ mile back to find his father Loren. Twin torments tortured his mind and pounded in his heart; the suffering of his favorite horse Vic, and the fear of his father's rebuke. Both of these anxieties were short lived. His dad was concerned, but understood. He did not blame Donald.

They hastened back to the site of the "harrowing"

experience. Working together they were able to lead Vic out of the stream unharmed. The horse got a well deserved rest. Donald's father worked with the broken horse harness in his times of rest and repair. Pride of his leather craft artistry made the labor at such times more relaxation and recreation than work. Horse, harness and horseman were all back in good shape.

If there is a hero in this harrowing happening, it must be old Vic. On a distant day in a place of the convergence of horses and men, we may get old Vic's take on the tale.

When no one holds the reins

Don G.'s father Robert thought it safe enough to leave the horses unattended and hitched to the corn binder. He let them rest in the field while checking on things in the house nearby. It appeared not to be such a good move when Robert looked up and saw the team apparently spooked by something, run off pulling the unmanned corn binder behind. With increased speed came increased noise. The horses knew something was chasing them. That something seemed to be gaining on them. They ran through an open gate with the binder banging along behind. On they went until coming up to two trees. On seeing the trees the horses slowed down a bit as if uncertain whether to go around the trees or between them. When they chose to squeeze between the trees, the binder wedged enough to pull the horses to a halt. Wonder of wonders, there was no serious damage to the horses, binder or trees. Praise The Lord! Some days are like that.

Let's get out of here—now!

Think about it. What would you do if you were a horse and this happened to you? Working with horses, you will get along much better if you have some "Horse Sense." Horses know you can't trust one of those tractor type machines. One of those noisy things is not a horse no matter how loud it boasts of its so called "Horse Power." Jeff H. was loading the manure spreader using the skid loader, (a big motor powered pitchfork). The horses, Ed and Fred were hitched to the manure spreader. They were unattended while Jeff was operating the skid loader to pick up the manure in the barn and load it in the spreader. He knew they were a little nervous that close to the skid loader with the motor running. When the horses jerked he called them down and they stopped. Jeff thought he'd wait a little so they would get used to the noise of the skid loader. After getting off the laborsaving loader, he picked up the pitchfork to throw some manure in the spreader by hand. The horses were used to him being around them with the pitchfork. Jeff thought this would help to calm them. Just when you think you know what their thinking, you don't. They took off to the security of their home base in the barn with the spreader following close behind. No damage was done to the horses, barn or spreader. It was just a minor blow to Jeff's ego.

A few days later Jeff attempted to cut down a nearby tree. As the tree cut from its stump was falling, it lodged in a tree next to it. Jeff thought pulling the tree out would be a good job for his horses. With a chain tied to the bottom of the tree he planned to have the horses pull it free. All went well as the horses started to pull the tree free, but Jeff and the horses were in for another surprise. The tree had a trick up its leaves. Branches and leaves came down with a swish. Hearing the sudden strange sound the horses took off through the yard, knocked over the mailbox and ran again back into the barn pulling the tree behind them. Teresa thought Jeff had decided to cut a tree near the barn, not near the house. She also was among the surprised when she saw the horses, tree and damaged mail box. The horses had no thought of the fine involved when a box for the U.S. Mail is deliberately damaged. It's probably Jeff's expense, but he doesn't need to tell the horses. Let them think about it for a while. Probably the horses couldn't care less.

You did what!

Mary Emma H.'s dad Hervey was at times a horse trader. Like school boys with marbles, horse trading gets into your blood. For some it was a profession, for some a hobby and others a good way to improve their herd of horses. Horse trading was always a risky business. At times even the best horse traders got taken in a bad trade. Often it was a spur of the moment thing. Taking time to think about it or talk about it with your family was just an excuse to back off from a slick trader. When you didn't step back and take a second look, you risked a bad trade and catching it from your wife or family.

Mary Emma's brother was out plowing with a well matched team of horses. Some people said he was a bit young to be doing a man's work, but that didn't bother Wilford. On such a warm wonderful day in early April it was great to be out in the fresh air working with horses. This team suited him. The horses worked well together. The sun warming his back felt good. Wilford was as satisfied as the birds that raced after the plow for worms and grubs as the fresh turned black sod was exposed to the light of day.

At the end of the field near the barn, Wilford and the team were resting when an old flatbed truck with high cattle racks appeared in the barn lot. Out of the truck cab stepped his father Hervey and another man Wilford did not recognize. The other man stayed by the truck as Wilford's dad walked over. Wilford was proud of the quality and amount of plowing that he had accomplished. He thought perhaps his dad had come over to admire his work. From the look on his dad's face he knew praise of his labors was not on his father's mind.

"Unhitch the team and take them to the barn," ordered his father.

"Why do you want me to do that?" asked Wilford, a bit confused and agitated.

"I want to take off the harness and put on the

halters," replied his father.

"The plowing isn't done dad. You aren't selling this team are you?" Wilford shouted back.

He knew the answer and he knew there was no use waiting for his dad to reply. Wilford walked away shaking his head in disgust. To say her brother was upset over the loss of his favorite team was a gross understatement. According to Mary Emma's memory, Wilford was mad!

Apparently their father got the message. The next day Hervey showed up with another team of horses to pull the plow.

Horse handling 101—putting up hay in the barn

There is a lot of growing up between, "No, you are not old enough to go into the barn lot," and, "Yes, you may use the car tonight." Boundaries were passed when it was safe to do so. Children were housebound until they could walk. A picket fence kept them in the yard, until it was safe for them to go beyond. Around the barn lot was another fence that kept the livestock in and the children out. When they had chores that involved going into the barn lot, children were trusted not to get in harms way with the livestock or the machinery. The first step into training for adult responsibility came with a pony to ride or an animal to care for. Somewhere in what seemed to be endless years for most family farm boys and some girls, a time came when they were needed to play the role of an adult. On farms where hay was grown, children who were old enough often rode the horse that pulled the rope which lifted the hay into the mow. What a memorable day for those who were granted this opportunity to adult responsibility!

In the 1930s on many farms hay in the field was loaded on a wagon by hand with a "pitchfork." A loaded wagon of hay was pulled by a team of horses to a spot near or in the barn where the loose hay was lifted up a bunch at a time by means of a metal harpoon or slings. These hay bundle hooks had a strong rope attached which ran through a system of pulleys. Bundles of hay were pulled straight up from the wagon to lock in a wheeled carrier on a track that ran horizontally the full length of the highest peak of the barn. The rope, about an inch in diameter, ran up to the carrier, through a pulley down to the barn floor through another pulley. The end of the rope was attached to a singletree which was pulled along the ground away from the barn by a horse. The hardest part for the horse was the initial pull which raised the bundle to the track at the barn peak. When the hay bundle reached the carrier, it rolled horizontally along the track with very little pull from the horse.

There's nothing to it—right?

David H. learned early on there was more to the job than a leisure ride on a gentle horse. It looked easy when someone else did it, but he found it to be a bit more demanding. His salvation was that the horse had been down this path many times before. It's not easy to fool a horse who knows who is in charge. It was quite a thrill to feel the horse get busy and push into the neck collar as the taut rope pulled the hay bundle off the wagon. The bundle was lifted skyward toward the waiting carrier on the track above. The hardest part of the pull comes just before the hay bundle reaches the carrier. Once a bundle locks into the carrier with a snap, the rope loosens and the pull is over. The horse knows an extra effort is needed to gain the relaxed rope. At times the extra effort does not come without the encouragement of the rider. As David discovered when a horse that does not give the extra push so the bundle will lock into the carrier, a young horse driver is in for excited counsel from an adult. It is not a welcome sight to see the hay bundle start back down as the horse begins to relax and even back up a bit. With a jerk of the reins David urged the horse to hit it again. When he'd see the rope slackened as the bundle was locked into the track carrier and hear the sound of its roll along the track, David knew it was time to turn the horse around and return for the next skyward trip of another hay bundle. On the return trip to the starting point the rope follows the horse. He learned from experience the horse had to be turned around the same way each time or the rope would end up in a twisted, tangled mess. Even though he finally managed to drive the horse like a pro, David preferred work with machinery. When he had a choice of using horses or tractor, the horses were out of the running.

The art of leaving a horse in a hurry

Many boys who mastered the art of riding a horse that pulled the hay rope were confident this was no job for a girl. Virginia B. proved them wrong. Some people were locked into the notion that a farm girl's place was in the house or the garden. They would surely have wagged their tongues about Virginia's labor on the farm that day. She not only was a girl, she also rode the horse in style, side-saddle of all things! Even though she was a young girl that rode a horse side-saddle, Virginia had proven she could do the job.

One afternoon riding the horse proved to be rather exciting. The whole process of putting hay into the barn mow came to a temporary but memorable halt. Virginia got the immediate attention of the men around the wagon loaded with hay as well as those forking the hay in the mow.

At Virginia's urging with a tug of the reins the horse moved forward pushing against the harness collar. With this forward thrust of horse power came a tightening of the traces back to the stout wooden singletree. As the horse moved forward and the rope tightened, the singletree came up off the ground. Thus began the tug of war between the heavy bundle of hay at rest on the wagon and the power of the horse. When the rope tightened against the tug of the hay, as had been proven time and time again, the resistance of the resting hay was no match for Virginia and the horse.

The hay bundle had been raised only about five feet off the wagon when it happened. Suddenly the hay bundle fell back on the wagon and the horse lurched forward with no pull of the rope to hold it back. For some reason the rope suddenly separated from the singletree. Usually when a singletree comes to rest it is pulled on the ground by the rope. In this instance, the singletree also taken by surprise, dropped to the ground and bounced back up striking the back legs of the forward moving horse. Virginia, now well aware that this trip was a bit unusual, tried to encourage the horse to stop and think about it. The horse did not see it her way. After all it wasn't her legs being hit by the singletree. The singletree seemed bent on hitting the horse on the hind legs and the horse tried to even the score by kicking back. Fearing the worst, men ran to help Virginia with her ride on a frightened horse on the run. Before they were able to get to Virginia and the horse, she did the thing a lady would do. Virginia took the side-saddle slip from the horse and singletree landing safely on her feet unharmed. The men worried about Virginia; she was worried about the horse. All you young guys who aspire to a Hollywood style rear mount of a horse don't stand a chance competing with Virginia and her side-saddle dismount. The horse stopped soon after Virginia slipped off its back. When the rope was tied securely to the singletree, work continued as planned.

On changing horses on dry land

Susanne K. had a pony named Betty which she rode to bring the cows in from the pasture some distance from the barn. The pony was better than her dog in rounding up the cows. Susanne loved to be out with her dad and enjoyed the horses as well. She rode "Old Fred" one of her father's horses to pull hay up into the mow. The horse was uncomfortable to ride. It was hot and humid. Her legs were sweaty. The backbone of the horse was not comfortable. Susanne thought of a way to fix the problem. She had a saddle for riding her pony. On her own she figured a way to put a strap on the belly-band attached to her pony saddle. With the pony saddle on the draft horse, she was truly up on her "high horse." No longer did she suffer from the sharp backbone and sweat of the horse. Now it was fun to be helpful putting up hay.

After awhile however, the routine became boring. Old Fred had reached the age when nothing much altered his slow deliberate pace of getting the job done. Susanne told her dad she would like to try riding one of their other horses that had more spirit. She wanted to try using "Star," a large beautiful black Percheron. Her father did not like the idea. He felt Susanne would not be able to handle him. Susanne kept insisting until her father, against his better judgment, agreed to let her try. The horse proved to be fast and strong pulling with determination. It was hard for Susanne to get him to stop and turn around to start for another pull. The horse seemed to be bored with this dull routine of little challenge. After an afternoon struggling with Star, Susanne was worn out and content to go back to Old Fred. It is hard at times to admit that "father knows best."

Not every boy can lead a horse

Bob Mc. did not get to ride a horse when it pulled the hay rope. He had to walk and lead the horse. It would have been much more fun riding on top of the horse than walking in front leading it. Bob had a hard time keeping the horse going just before the loaded bundle of hay would hit the "trip." At this point the rope slackens as the carrier runs on it's own along the track at the peak of the barn just under the roof. Most horses could feel the rope go slack upon hitting the "trip" and stop. This horse had to be led. So that's what Bob did, even though he really wanted to ride instead of walk. It might have been some consolation for Bob to know some people can't even lead a horse to water, let alone make him drink.

"Oh, no you don't!" "Oh, yes I do!" And she did.

"Why can't I go with Donald to ride a horse," whined Treva, Donald D.'s sister. "I could ride old Vic. Donald says he's gentle and wouldn't hurt me," Treva reasoned.

"I know old Vic is gentle, and would not mean to hurt you, but he is blind and you are just not old enough to guide him," replied her mother Ruby.

"Please mother, Donald would go with me. I promise I'll be very careful," her daughter persisted.

"Treva dear, I know it is hard to wait until you are old enough to ride the horses, but it won't be long. Right now I need you to help me get dinner ready for the men who will be in soon," requested her mother in the vain hope of at least a temporary closure on the matter. A dark cloud of disappointment drained the sparkle from Treva's eyes. She spent the rest of the morning doing what was expected with little talk and less enthusiasm.

At breakfast the next morning, Treva's father Loren announced that he would be gone most of the day to an auction. Even though she dearly loved to be with her father, she felt certain a request to go along with him on this trip would be denied. It was not worth the risk of another disappointment. After her father left and she'd finished her usual household chores, Treva spent the rest of the morning outdoors playing with sticks and stones making an imaginary pen for an imaginary horse. She did not see her father drive up with the old farm truck equipped with livestock racks to bring back any farm animals to be sold "worth the money."

Her father was in the house before she knew of his return. "Oh daddy, you're back!" she squealed. He picked her up holding her gently in his strong arms and said, "I have a surprise."

"Tell me daddy. Tell me what it is," Treva shouted with excitement.

"I told you it was a surprise. Close your eyes. Put

your hands over them and wait until I tell you to look. No fair peaking, OK?" asked her father.

A definite "OK" was her excited response. Donald had been instructed to bring the purchase of surprise into the yard. When her father saw Donald enter the yard, he told Treva and her mother to come out on the porch and take a look. The balloon of Treva's anticipation exploded into ecstasy at the sight of a pony. Her squeals of delight were ample reward for the sacrifice both parents made for the center piece of family recreation.

"A pony! Oh Daddy, a beautiful black and white pony," was her response quickly followed by, "Can I ride him now?"

Trying not to dampen her enthusiasm he cautioned, "The pony needs to get to know us before we ask him for a ride. Let's go over slowly. We don't want to scare him."

"Daddy, what's his name? How can I talk to him if I don't know his name? He is a boy pony isn't he," asked Treva?

"Yes, he is a boy, a grown-up boy pony," said her father. "He won't get much bigger. I didn't ask the man who sold him to us about his name. Since the pony has a new family and a new home, why don't we give him a new name? Is that alright with you ladies," her father asked as he slipped a wink to Treva's mother?

Her mother said, "We had a pony and his name was King. I don't think that's right for this pony. He should have a special name. What about a name that rhymes with pony? Do you have any good ideas Treva?"

Treva puzzled over the matter until a silly word came into her head, "Bony," she said with a giggle. The words "bony pony" had everybody laughing for the fat little pony was not bony. "What about Tony?" suggested Donald.

"Tony, Tony," Treva said trying to get a feel for the word on her tongue. Then, as if the full impact of a new chapter in her life shone brightly before her, she said, "Tony the pony, Tony the pony, Tony the pony. That's a good name for him. Now can we go up and tell him," she asked?

Tony the pony had a new name, new family, new home and a very special friend. Treva loved and challenged that little pony and Tony the pony reciprocated. Up and down the lane in front of the house she rode. But Tony soon tired of this going and going and not getting anywhere. It came as quite a surprise to Treva the first time it happened. Perhaps not so much a thing that happened as what Tony did when he'd had his fill of this back and forth business. With a sudden stop, off went Treva over the lowered head of the pony. Her mother saw it happen and started to rush to her daughter's aid when Treva got up and climbed back upon the pony. Treva's mother wisely waited to allow her daughter to come to her own terms with the pony. In the days to come this little game of who's in charge was repeated over and over again. A very plucky young girl cultivated coping skills equipping her for life. Tony the pony did indeed have a very special friend.

Just what Tony needed—a mother

Treva, Donald D.'s sister had taken over ownership of Tony the pony. It came as something of a surprise when another claimed the rights of mothering. When the family stopped to think about it, Mae, the mare was the logical one for the job. Mae, much like Tony, came from outside the family circle of horses. When you are an outsider, survival depends on your ability to find or make a place of belonging. Soon after Tony the pony arrived on the farm of this special family, Mae discovered her place. Mae, "the mothering mare," took over when Treva returned her pony to the horse paddock. Tony the pony did not suffer the nickering censure of the larger horses while Mae was around. It did not take the pony long to realize he was in the protective custody of an equine angel. Along with the other horses, Tony had his place in the horse barn at night and during the day in times of weather extremes. The horses were tethered each in its own stall. Each horse had a substantial wall between it and the next horse. Some horses were inclined to kick at times. The wall between kept them from the injury of another's kicks. Tony was the only one without a stall. He was tied in an open area near the other horses. However, like the others, he had straw or corn fodder bedding behind his wooden feeding box.

Every evening before going to the house for their supper, Treva's father, Loren or brother Donald fed the horses ear corn, (corn on the cob), a few oats and hay in a feeding box in the front of every stall. The pony lived on grass, oats and a little hay. One morning while checking on the horses, Donald noticed a half-eaten ear of corn in Tony's feeding box. He wondered how that half-eaten ear of corn had gotten there. He was certain his father had not put it there since the pony did not need the corn fed to hard working horses. Rats around the barn ate some of the ear corn, but they usually did not bother the horses. It was a puzzle. While making his rounds among the horses, Donald discovered Mae "the mothering mare" was not tied in her stall. They may have forgotten to tie her the night before or failed to make the rope knot secure. After tying Mae's rope to the stall and giving her an affectionate pat on the rump, Donald walked back past Tony. Remembering the half-eaten ear of corn in the pony's feed box, he thought to himself, "Can it be old Mae carried an ear of corn to the pony?" That evening Donald and his father fed the horses as usual, except they left the mare Mae untied. Sitting down on a bale of hay where they could see the pony, they were witness to what they had suspected. Mae, "the mothering mare" carried her golden gift of an ear of corn to the charge of her choice, Tony the pony. Donald and his father sat in silent awe as they witnessed "Horse Sense" compounded with the compassionate mothering of a foster mare.

Tractors Take Their Turn
Farming Changes With Tractors

Between 1837 and 1959 farm equipment underwent more change than in the previous history of mankind. All over the world enterprising inventors and manufacturers were making tremendous improvements in the equipment farmers had used for thousands of years. During this period farmers got off their feet and rode on the equipment with which they farmed. The principle power source changed from animals to the internal combustion engine.
Boyd C. Bartlet—*John Deere Tractors and Equipment:*
American Society of Agricultural Engineers

Our generation had the opportunity to stand on the narrow headland between the furrows in time of hand/horse labor and tractor mechanization. Some of us maintained a love affair with horses and ponies; others have been captivated by the hands-on power potential of the tractor and farm machinery.

Transition to tractor power

It could have been fatal. I am fully aware of that. I don't blame the John Deere Model A tractor for the accident. As a 13 year old with five years of tractor driving experience, I should have known enough to slow it down before turning the corner at full throttle with my foot firmly planted on the left wheel brake. After all, this new tricycle type tractor had been on the farm for three days. Any bright farm boy who claimed to be tractor wise should have known all about it in three days. I didn't. The tractor threw me off and landed on its side with my left leg pinned under it. No bones were broken, but it took over five years for the flesh to recover.

Later that summer, I climbed back on the tractor with a bandaged leg, not as smart as I thought before the accident, but now with greater caution and wisdom I continued my tractor driving adventure. I'm on the tractor side of the great divide, but with the size, complexity and cost of tractors today I suggest we take another look at the horse.

Twice told tale of the horse/tractor separation

Mary Eleanor H. had two uncles. One uncle had a team of horses. The other had a tractor with steel wheels. She remembers the uncle with the horses plowing a garden with a walk-behind, hand-guided, single-shear plow. The uncle with the steel-wheeled tractor pulled a plow on wheels with two shears. On the rear wheels of the tractor were steel cleats that dug into the ground. The marks of the cleats were greater in number, deeper and more uniform than those from the shoes on the horse. Horse and tractor, each left their defining mark in the soil of the earth and the soul of man. One uncle turned off the lifeless tractor at the end of the day, satisfied with the greater gain of plowed ground. The other uncle ended the day of labor with the usual horse and man routine ritual. He relieved the horses of the burden of the harness and wiped away the collar sweat from the necks of the gentle giants. Only after he'd watered and fed his horses did he think about his own rest and repair. While the identifying marks of cleats and feet were soon lost in the seeding of the soil, the imprint difference remains in the heart of man.

"Once upon a time there were two brothers," is the open door to a multitude of stories of character differentiation. This story played out in our area has a brother on each side of the horse/tractor predilection. Jim, David and Becky H.'s uncle Howard was one of the first farmers in Clinton County to own a tractor and their father Bob was one of the last. Howard did not care for horses and Bob did not want to do without them. The tractor for Howard was the laborsaving way of the future and vital to progress in agriculture. Bob farmed with horses as long as he could. He sowed wheat to get an early start before the corn was ready for harvest. Bob did it with a forty-inch grain drill. What pulled the drill between the rows of corn? Not a tractor, it was a horse, of course. He shucked corn by hand and threw it into a wagon drawn by a team of horses. Bob made the change to the use of a tractor only because he could not get repair parts for broken horse-drawn equipment. Bob had only two teams at a time, but loved his horses.

Let the horses out—throw away the lock

Even though he finally managed to get it together and guide the horse like a pro, David H. liked to work with machinery. When he had a choice of working with horse or tractor, the horse was left in the pasture. His first tractor was an International Farmall F12 purchased primarily to replace the horses plowing/cultivating corn. One of the pluses for the tractor was a mechanical lift. With it there was no more of the "arm-strong" labor required to raise the weeding shovels of the horse drawn cultivator. At the fall Corn Festival as you might expect, David was driving the International, not a team of horses.

The transformation from horse power to tractor power for Ray C. and his father came in the 1930s. At the time they had 4 or 5 draft horses. A steel-wheeled International 10-20 tractor was purchased for plowing, disking and belt power.

Their first rubber-tired tractor was an International Farmall H, bought in 1941. In 1943 their last team of horses was treated to a truck ride off the farm to a family who had greater need of horses.

One might expect a person who loved to work

with horses like Donald D. to find fault with their tractor, a steel-wheeled Fordson. The tractor was used to pull the double disk but it didn't have a lot of power. When turning a corner pulling the planter in loose soil, the front wheels had a tendency to skid instead of making the turn. It was much easier for the horses to make the turn in these conditions. Even driving straight it was hard to steer. Starting the Fordson was something else. It was very hard to start in cold weather. Donald found about the only way to get it started when it was cold was to drain the oil and heat it, then put the heated oil back in the tractor. This tractor might have its place, but a horse it was not.

Tractors take over

Tractor manufacturers and engineers heard the farmer's complaints. Tractors were modified by putting the front wheels close together with rubber tires and wheel brakes so they were more maneuverable. Tractors soon became better equipped to take the place of horses. Sharon G. drove a tractor to pull up hay bales into the haymow of the barn. They used the same system being used for putting loose hay in the haymow. When the heavy load of hay was pulled to the top of the barn and locked into the carrier on a horizontal track, it didn't take as much power to pull the rope. Since the wisdom of a horse who knew when to stop pulling so hard could not be transferred to the tractor, it was up to the driver to watch for the slack in the rope and stop the tractor. The tractor driver who failed to stop soon enough was certain to hear from the guys in the haymow. It was customary to put the tractor in reverse and go straight back to the starting point. Trying to turn around as was done with a horse, was not acceptable. The man pulling the rope back into position was not too fond of trying to pull the tractor back or the rope from under the tractor wheels when the driver backed over the rope. Sharon was quick to learn the do's and don'ts of using a tractor when you didn't have a horse to blame.

Let tractors do their thing

Mike D.'s grandfather Frank's first tractor was a two-plow International with steel wheels. In winter he used it for plowing. The weather was bitter cold so he started the tractor pulling the plow and let it go slowly on its own guided by the left wheels in the furrow. His grandpa stepped off and walked beside the moving tractor to get warm. When it came to the end of the field, Frank would get back on and guide the tractor into the furrow going back the other way. Then as before, he'd get off again and walk. He kept doing this until he was warm enough to ride for a while. Little did Frank know his grandson would one day watch a tractor-type Moon Rover moving on the moon with no one on board.

Susanne K. liked to help her father Bob on the farm. Other than riding the horse to pull up hay into the loft, she did not work the horses. Her father used a team of horses to pull the corn planter that checked their corn. Susanne helped her father move the corn check wire over at the end of each row. (See footnote). They later got a Ford tractor and used it in place of the horses. Not long after, they had only one team of horses left on the farm. Susanne worked with a tractor in the field in preparation for planting after her father bought an International Farmall M with rubber tires.

Corn check wire was strung out the full length of the field to be planted. It was made with small knobs spaced equally along the wire forty inches apart. With the wire in place on a forked trip lever on the corn planter, three or four corn kernels were dropped the same distance apart for the whole field. This checker board effect made it possible to plow out the weeds both the length and width of the field. At the end of each row the corn check wire had to be moved over for the next row all across the field.

Risks on the road to responsibility

Even before tractors and farm machinery, farm families were aware of the ever present potential for a life threatening tragedy. Growing up on the farm involved a continuing confrontation between the blessing of responsibility and the curse of the risk. It is always a tough call for parents. There is no fool proof detour around the risk of accident on the road to responsibility. Every accident can be fatal or a learning experience. The enthusiasm of family farm children to take responsibility is gratifying, but it is often subject to risk. Most of those interviewed had memories of personal and neighborhood tragedies, better kept as lessons of learning for living rather than festering wounds of guilt and regret.

An early start

Before he was old enough to go to school, Tom W. drove a steel-wheeled 10/20 McCormick Deering tractor. He really didn't drive it, but Tom sure thought he did. As a young boy, Tom took a ride with his father Tommy on the tractor while it was pulling a plow. The tractor did not go very fast. With the wheels on one side both front and back in the furrow, the tractor guided itself. Tom's dad stepped off the tractor and walked slowly beside his son in the seat by himself. Tom remembers thinking how great he was driving that big tractor. It didn't last long. His dad quickly got back on the tractor with young Tom still sitting on the seat protected between his dad's legs. Many a farm boy got a very early start on a love affair with farm machinery, especially the tractor.

On the farm until Tom was 12 years old, they had both horses and a tractor. He did not harness a horse, but his father taught him how to drive a team of horses. For some reason, that did not appeal to Tom nearly as much as learning to drive a tractor.

On taking out a little corn with the weeds

The first tractor on Don G.'s family farm was an International Farmall F20. The first field work he did with this tractor was to plow/cultivate corn. It is not a job for beginners. As might be expected, it takes constant concentration to keep from plowing up some corn as well as weeds. Cultivating corn with a tractor meant driving between two rows of corn. Shovels on each side of the corn rows were designed to plow out the weeds and allow the corn plants to slip between the destructive shovels. The temptation with this arrangement is to try to do the impossible, look at both corn rows at the same time. In spite of being told it is necessary to watch only one row at a time, a person cultivating corn for the first time seems compelled to do otherwise. There is some subversive compact between the hands and the eyes which trick you to turn the tractor at the same time you take a quick glance at the other side. With just a slight turn of the steering wheel and zip, you have plowed out a little corn with the weeds. When you glance back to the row you were supposed to be watching, the damage is done there too. At that point you really get into trouble, because you are unable to resist the urge to look behind to see if you actually did plow up corn in both rows. In the time it took for that backward glance, you have steered the tractor off course enough to take out three or four more feet of corn. Experience is the great teacher. You will learn not to do that again.

The tractor refused to take the blame when Don managed to plow out some corn along with the weeds. When his dad saw the uprooted corn he was not too happy. Actually he was not happy at all. With the experience behind him, Don learned the art of taking out the weeds and not the corn. Once he had mastered the art of keeping his eyes on a single corn row, he was challenged by his nodding head in need of a nap, but that's another story.

Time to take a stand

Their first John Deere row crop tractor was a Model A. The cultivators bolted to both sides of the tractor had to be raised and lowered by hand. Tom W. was young and could not raise or lower the cultivators while sitting on the tractor seat. His only alternative was to stand up and use both hands to push the lever forward that lifted the cultivators up above the ground. This struggle had to be repeated at the end of every row. Boys larger and stronger were able to raise the levers on each side of the tractor while it was still moving. The maneuver at the end of the row added to the problem. At this point he had to turn the tractor around and drop the cultivator shovels back into the ground for the return on the next row. At first all Tom could do was to stop the tractor at the end and raise both levers before turning around. It was not as hard to lower the cultivators without standing up or stopping the tractor. He hoped the tractor would not get out of line and plow up the corn. As Tom remembers, he did not plow up too much corn. Cultivating or plowing corn, no matter what you call it, was a challenge for young farmers. In time Tom found he too could raise the cultivators, turn the corner and lower them with no need to stop the forward progress of the tractor. Not long after he had learned how to handle this hard part of the job, tractors were made with laborsaving mechanical and hydraulic lifts. Then a farm boy could cultivate corn with the best of them.

Note—I'd challenge any adult who has never plowed corn to try if they think they could do as well as the boys. But you are off the hook because very few farmers plow/cultivate corn these days.

When out of sight was not good enough

For a young man and some older men, the roar of the tractor motor speaks of power and is music to the ears. Because of this, there are times when one would be better served to use a team of horses. For example, Bobby Q. a young farmer confident of his skills as a tractor operator was cultivating corn away from the farmstead. Usually corn is cultivated with the tractor throttled down at a slower pace to be careful not to get too close to the corn and take it out with the weeds. After a while this slow boring pace gets to a young man. Why not open the throttle up a bit to keep from going to sleep? Bobby knew his dad was out of sight and figured he'd never know the difference. After a round or two of increased speed, he could not resist "hot-rodding" and pushed it to the limit. The roar of the engine was pleasing to hear. Bobby did not think about sound traveling in ways that sight could not. His dad Frank knew what he was doing from the sound of the tractor. The sound was not quite as pleasing to his ears and he came out to the field to put an end to Bobby's fun. After a few years of experience, Bobby learned even the seasoned tractor drivers plowed out some corn with the weeds if they drove too fast.

I've had it—I'm out of here!

Dick G. was plowing one spring with their Allis Chalmers WD tractor just after dark. He was driving along with everything working as the saying goes, "without a hitch." The tractor was purring along hitting on all four cylinders. Although it had gotten too dark to see the Oliver two-bottom plow, Dick felt all was going well. Even though he was not yet an adult, he was confident he was as capable as most adult tractor drivers. He had learned to hear the different sound of the tractor motor when it was easy going in light soil or pulling down in clay packed "gumbo." Suddenly, there was a loud bang like the shot of a gun followed by a forward lurch of the tractor. Even though it came in the darkness as a total surprise, Dick knew what had happened. The plow had hit an immovable object like a large rock. The hookup between the tractor and the plow has a break-away safety device that breaks the connection between tractor and plow. The clevis release is designed to prevent damage to the plow. In the light of day it would come as a surprise as well, but after dark it seemed much more ominous.

Since there was no backup light on the tractor it was hard to see the abandoned plow. Staying with the left front and back tractor wheels in the plow furrow, he slowly backed up until he saw the plow. Dick was about in position to make the hookup between the two when the tractor motor quit. In the quiet and deepening darkness, Dick knew he had gone one round too many and the tractor was out of gas. What to do? It was not exactly a "win/win" situation. His dad planned to come out and take his turn plowing, but Dick decided not to wait until he arrived.

Without the benefit of a flashlight or a bright moon, he began to walk across the unplowed ground rather than stumble over the soft unstable freshly plowed soil. It seemed like the way to go until he came too close to the residence of a night resting covey of Quail. Quail prefer to be left alone day or night. When danger comes near, they do not take time for a counsel meeting or packing their things. Day or night they take off in an instant with a flurry of wings like the sound of a July 4th bottle rocket. Such a sudden surprise in the dark of night does a number on the nerves. When Dick saw his dad with flashlight in hand walking toward the tractor, he didn't even stop to give him an account of the twisted turn of events. What a way to end the day! Tomorrow was bound to be better.

Thank God for help when you really need it

His mother and dad were leaving for the weekend and his brother was gone for the day. Dick G. was by himself, or so he thought, when he decided to go out and mow hay. While mowing, the cutter bar caught on a clump of orchard grass that was too tough for the mower to cut. Usually when this happens the trip mechanism will release the bar so it disengages and swings back trailing the tractor and mower. For some reason this time the cutter bar malfunctioned and rose up behind Dick's back pinning him against the tractor steering wheel. He was in some pain, but the crucial problem was that he could not get out from under the heavy mower bar. No one was around to help him escape the vise grip of the mowing machine and the steering wheel. What could he do?

Dick did not know his parents were late leaving. His parents did not know he had plans to work in the hay. For some reason on their way to leave, they just happened to look toward the hay field and saw the tractor stopped in the field. It did not take them long to realize Dick had a serious problem. With his dad's help he managed to wiggle out of his entrapment. As Dick was being pulled out the weight of the bar fell on the steering wheel with enough force to bend it. Since he was fortunate not to be seriously hurt, his dad made it clear to Dick he was unhappy about the bent steering wheel. Sometimes it is hard for fathers to confront and confess their innermost mix of feelings of pride and anxiety concerning their children's labors with livestock and machinery.

If it weren't for the mud

It had been a wet fall and there were places in the hog lot that were soft enough for a tractor to get stuck in the mud pulling a wagon fully loaded with shelled corn. Before he left for the Master Mix mill, Bobby Q. had to wait for the arrival of the school bus. When his son Marty and friend John got off the bus, he told the boys about the mud. The boys probably would not get the tractor stuck if they unloaded some of the corn at the first feeder before going on to the other feeders where the mud was worse. Satisfied the boys understood, Bobby left for Wilmington with a truck load of corn to be ground for livestock feed.

It was late in the afternoon. Bobby Q. was at the Master Mix feed mill and grain elevator with a load of corn still on his truck when he got the call. As often happened there was a line of loaded trucks at the mill. He was passing the time visiting with other farmers while waiting their turn to have their trucks unloaded, when someone told Bobby he had an urgent call on the office phone from his wife, Wilma. An urgent call from a farmer's wife could be about any number of things, from livestock breaking out to someone getting hurt. The burst of anxiety ballooned to gut-wrenching fear when Wilma gave him the bad news.

Both boys in their early teens had tractor driving

experience. Marty was going to drive the tractor but when John said he wanted to drive it, Marty agreed. For some reason the boys were distracted and drove right by the first feeder. Then it was too late to stop so Marty suggested they go on to the next feeder with the wagon still full of corn. At the second feeder they were able to unload some of the corn, but at the third, the tractor got stuck in the mud. And then it happened. When the boys tried to raise the tractor out of the mud, the front of the tractor rose up and turned over backwards. Marty was able to jump free of the tractor. John sitting in the seat driving the tractor was unable to jump away. The tractor came over on top of him.

A friend insisted Bobby take his empty truck and hurry back to the farm when he learned of the emergency. Racing back to the farm Bobby feared the worst. The tractor had turned over on the boys was all he could think about. The torrent of tears made it difficult to see the road clearly at times, but Bobby was home in record time. The sight of the overturned tractor with all four wheels up in the air was so devastating he seemed almost unable to grasp the significance of the sight of both the boys standing safely in the yard. Nor did he notice at first both boys were barefoot.

Wilma had told Bobby the boys were okay when she called the mill, but it just didn't register with him at the time. After the wave of relief swept over him, Bobby was told of the miraculous rescue. The accident happened because of the mud and because of the mud neither boy was hurt. Marty's friend John was pushed unhurt into the mud by the same miraculous power that made it possible for Marty to pull him out. The mud gave the boys a reminder of the need for caution when working with farm machinery. Four boots, shoes and socks were claimed by the mud until they, like the tractor, were retrieved to be put back to work.

Father's turn to fly low

Bobby Q. was not certain why he went to check on his son. Marty had gone with the tractor, disc and harrow to work ground in preparation for planting. The field near New Vienna, Ohio was some distance away from their farm. Bobby just had the feeling it might be a good idea to check and see how it was going. It turned out there was good reason for checking. Bobby discovered his son had a badly bleeding hand. Just minutes before a harrow slipped out of his hand while Marty was unloading it. As it slipped a sharp part of the harrow cut a deep gash in his hand. Marty obviously needed to see a doctor right away. Bobby suggested they take Marty's car. It had power and potential for speed to get them there in a hurry. Driving to get his son to the hospital as soon as possible, Bobby pushed the accelerator to the floor. In contrast to driving his truck Bobby, was roaring down the road at 95 m.p.h. and the Pontiac Trans Am still had power to spare. At this point Marty put in a bid for a slower pace saying, "Dad, I'm not hurting that bad!"

With one wrong turn of the steering wheel

Most of us do not back up with as much ease as we go forward. It is troubling because backing up is often an admission of a mistake. Obviously we gain no ground if we spend more time and energy backing up than going forward. However, it is just as clear we need to back up and change our course or try again when we have made a mistake. With farm machinery, at times we must back up and make another run for it. We have a problem when driving a tractor or truck pulling a wagon or another piece of mobile equipment. It is hard enough to back up in a straight line, but the real challenge is to back a piece of mobile equipment around a corner. If nothing is being pulled by a tractor, car or truck, the turn of the steering wheel is in the same direction going forward or backward. Turn the steering wheel to the right and the vehicle goes right. It is the opposite if you are pulling a tongue-drawn vehicle. Ray C. found this to be the case in a moment of panic when he was 12 years old.

Many farmers with hay to harvest did not have a bailer of their own, when Ray was young. Farmers who invested in a bailer met some of the expense of their investment by doing custom work for those who had none. It takes at least two people to operate the bailer. Ray's father Hower C. had both the bailer and two sons to help. On the morning that Ray had his tractor backing skills put to the test, he had no reason to doubt his ability to drive a tractor.

Ray drove the tractor and his dad and brother sat on the parallel platforms at the rear of the bailer. Their job was to secure the two wires that tied each bale, as well as oversee the operation of the bailer. The hayfield of the neighbor they were to do this day was new to them. It was obvious this field would challenge their bailing skills. The first windrow of raked hay hugged a deep ravine with no fence between it and the hay to be bailed. All went well as men and machine picked up the loose hay and bound it into two-wire 70 pound bales. It went well in spite of Ray's growing fear he might drive too close to the chasm edge only a few feet to their left. He was certainly relieved when they came to the corner and turned away from that ditch which seemed to Ray as deep as the Grand Canyon.

Ray felt he was home free as the tractor and baling crew started the rather steep hill climb up from the valley of the deep. With confidence in the power of the tractor to take the hill climb, the bailing continued. Nervousness was on the rise as the whining pitch of the tractor motor slowed to more of a groan. Ray's grip tightened on the steering wheel. His mind was still on the deep ravine behind them. A bit rattled, Ray failed to shift the tractor to a lower gear. With a groan, growl, sputter and cough the tractor motor was silent. Under normal circumstances this would be a time to rest a bit. With the silence of the lifeless tractor, the guys might savor the sounds of nature around them.

This was not to be since the silenced tractor and bailer were slowly slipping backwards down the hill. It looked like it could be a rough ride and fatal fall into the bottom of the deep ditch. The simple solution of turning the tractor and bailer away from the dangerous ditch came to Ray in a flash. No doubt he had frantic words of instruction from the two guys on the bailer. All it would take would be a careful left turn of the steering wheel. This maneuver would bring the tractor and bailer to rest on the side of the hill parallel to the ditch. Impulse overruled experience and Ray turned the steering wheel to the right which was obviously wrong. Ray's father and brother ended up still on the bailer just at the edge of the drop.

I hope you are not totally confused when I tell you, as Ray told me, the wrong turn of the steering wheel turned out to be the right thing to do. The left rear wheel of the tractor backed into the frame of the bailer and stopped the slide. Brother Ken and father Hower failed to appreciate their up-front view of the ravine. A lively discussion followed, but Ray did not go into that. Ken 11 years older than Ray, took over driving the tractor. Ray took the vacant seat across from his dad on the bailer. This was not just another day in the life of these bailers of the hay!

What's so funny?

Between triumph and tragedy often there is a minor player with a major role. We are not well served by either ignorance or anxiety of the outcome. Faith and trust alone will carry the day when our lives hang in the balance. Like the weakest link of a chain, the failure of one minor piece of equipment proved to be a major disruption of the day's plans for Ray's Dad Hower. Hower and the hired man Wilbur, were taking a load of hay on the road from the field to the barn. Hower driving the Minneapolis Moline tractor enjoyed the satisfaction of piloting the powerful machine. The hired man was taking a well deserved rest riding on the wagon loaded with hay. Neither man had any reason to believe this trip would be any different from the many other times they had used tractor and wagon to haul hay.

Between weight of the heavy load of hay and the pulling power of the tractor to move that load was a tough metal "hitch-pin." The pin was five or six inches long and less than an inch in diameter. Obviously I do not need to tell you why it's called a "hitch-pin." Few things follow a tractor without the little praised hitch-pin connecting the wagon tongue to the drawbar of the tractor. If you doubt its importance, ask any farmer who has spent valuable time looking for a misplaced hitch-pin.

Ray's dad and the hired man thought a lot about it that afternoon, after the worn pin dropped out breaking the connection between tractor and wagon. Hower jumped off the Minneapolis Moline tractor hoping to grab the wagon tongue and try to guide the wagon from going into the roadside ditch. However, as he jumped off the tractor he did not pull hard enough on the hand clutch control to stop it. The unattended tractor continued a slow, steady grind, about 10 m.p.h., down the road. Ray's dad was caught between the proverbial "rock and a hard place," between the "Minnie Moline" faithfully continuing to do its thing without proper guidance and the drifting wagon gradually coming to a halt with no power to keep it rolling.

It did not take him long to realize the wagon left on its own would soon quit moving, while the tractor kept up its steady misguided meandering. Ray's dad took off on foot charged with the adrenalin of seeing a very valuable tractor continue on its gear grinding journey into unfamiliar territory. The tractor had a head start going about the same speed as Ray's dad. He tried to get close enough to climb upon the tractor drawbar within reaching distance of the hand clutch control. A yank of the lever would stop the forward progress of the escaping tractor. Without so much of a backward glance, the tractor seemed determined to outdistance the man chasing it. Hower was only inches away just out of reach of the tractor when, it made an evasive turn into the side-ditch of the road. It crossed the ditch with a crash through the farm fence. The tractor proceeded at a slow persistent pace on its high adventure across the pasture lot.

Ray's dad had to break stride a bit going down into the side ditch and up through the tangled wires of the bedraggled fence. Across the pasture field he raced gaining on the moving Minnie Moline. It looked like he was about to win the race, when the unrestrained tractor took a sudden turn. This proved not to be a good move for the tractor. Straight ahead was an unsuspecting large tree. When it charged the tree head-on, the race was over. Bottom line—one damaged farm tractor and one very winded farmer.

Where was the hired man? Wilbur was still on the hay wagon with tears in his eyes. He was shaken up, but not for fear of the outcome. The whole scene unfolding before him turned out to be classic comedy. He shared the anxiety with Ray's father at the moment of indecision between guiding the wagon of hay and chasing the runaway tractor. However, watching the tractor chase had him doubled over with uncontrolled laughter. Hired men can become fired men. This thought did not occur to Wilbur at the time. His place with Ray and his dad was secure. Good hired hands are dependable, versatile, and stay with you making few demands because of their love of work on a family farm. Wilbur had found his place of service as a hired hand. Given time too cool off from the heat of the chase, Ray's Dad Hower, also saw the comedy in a near calamity.

We are blessed by the simple ties that bind us to the soil, to livestock, to others and to God. All too often these ties get little notice or appreciation. We take them for granted until they are not there when we need them. Are we losing the family farm because we neglected the hitch-pin of relationship? The wagon of cooperation is slowly drifting to a halt. As we sit watching on our own wagon, the God-given power of bonding to all creation continues to move farther and farther beyond our grasp. While the will and power to work together is still in sight, it's not too late to regain control before inevitable destruction on the tree of corporate greed.

Let There Be Light
Living Before and After Electricity

Electrification made all kinds of farm work easier and dramatically altered such enterprises as dairy and poultry production, facilitating mechanization and a stunning increase in efficiency. It also altered the rural home, allowing the introduction of a vast range of devices and conveniences urban people enjoyed.
David B. Danbom—*Born in the Country*

Many have vivid memories of the remarkable change between the dim lighting of kerosene lanterns and the light and power of rural electrification. Few people wish to go back to the darkness of those days. Yet we have great admiration for their dedication to each other, the livestock and the land, in spite of the darkness of their time. We have much to learn from their ingenuity and frugality. Of the many dramatic changes in the operation of the family farm in the years between 1930 and 1950, none were quite as immediate as electric light and power. The flip of a switch blew out the flame of kerosene lanterns. What had become routine to "city folks" a few years earlier, was a new exciting experience for most farm families. In some rare cases it was not easy to give up the old.

Are you sure that will work?

Scott K. shared a story he'd been told about a man who was a bit skeptical of the newly installed light bulb. He used it only briefly to help him see to put a match to his kerosene lantern. Once the faithful lantern was burning brightly, he turned off the electric light.

Press A. helped an old farmer friend who had never lived in a house with electricity move to a rental house where a few electric lights had been recently installed. The lights were both a worry and a wonder to the man. In this rental the bedroom was upstairs. A mattress has no desire to be moved and refuses to cooperate especially when being lugged up a flight of stairs. It was quite an ordeal for the old man and Press. Once up the stairs with the mattress in place on the bed, Press said, "I'm sure glad that's over."

"Yep" agreed the man, then added, "But there's another problem."

"What's that?" asked Press.

Looking up at a single light bulb on the ceiling of the room the old man remarked, "How'n the devil is a man going to blow that light out way up there?"

We won't know until we try it

Dan S. told me this story I have modified a bit. Before electricity came to their hill country, a man heard he could use a battery to light a bulb. When Josh told some of his friends about it they decided to try it. They had to walk to town, but they were able to get a used car battery and light bulb. When they returned to his home in the hills, Josh and his friends took two short wires and tied them to the battery terminals. When the wires were connected to the light bulb, it worked! It was something great, but Ed thought it did not give enough light. He had an idea for making the light brighter. Ed told the others he had to go to town, but would be back again soon. True to his word, Ed came walking in with a large roll of electrical wire. When the others asked him what he was going to do with the wire he said, "Give me a hand. I want to run two wires from here up to the top of the hill."

"Why are you doing that?" they asked.

"Just help me get it up there, and you'll see," he replied.

After the others had finished stringing two wires from the house to the top of the hill, Ed instructed, "Now go down and fetch me that batt'ry."

Again the others raised the question, "Why are we doing this?"

"This is the way I figure it." Ed told them. "You know, water picks up power when it runs down the mountain. Stands to reason it's the same with electricity, don't it?"

It sounded right to most of the others. They carried the battery to the top. After securing the wires to the battery terminals, the men walked down to see how much brighter the light might be. Some thought it better, others were not too sure.

One of the younger men figured there was a simple way to be certain. If electricity, like water did gain power coming down the hill, what would happen if they put the light at the top and brought the battery below. However, being young and bright, he decided to keep it to himself.

What made you think I was afraid in the dark?

Many who remember life before electric lights have vivid memories of kerosene lamps and flickering shadows. Boys who lived on dairy or beef cattle farms 60 -70 years ago, are now old men. Yet, they still remember the challenge of climbing a vertical ladder up into the haymow to pitch hay down for the cattle in the barn below. Imagine a young boy carrying a pitchfork and a kerosene lantern in one hand and using the other hand to climb the ladder. Add to the picture a cold winter evening in a drafty barn. Even though you were at the age when you wanted to be known as fearless, you were not. The fear of "who knows what," lingering in the flickering shadows, was backed by stories of tramps and gypsies who were reputed to sneak into a barn on a dark and windy night for a night's rest. These stories coupled with the potential for fire from the lantern, made for a hurried forking of the hay for cows in stalls below impatiently waiting to be fed. They had no heart for a frightened boy high above them frantically forking what was mistakenly called loose hay. On a dark night, un-baled hay is as determined to stay in the haymow as a boy is to leave it. Neither the resting hay nor hungry cows cared that his heart was pounding at the frightening creak and groan sounds of the overhead rafters.

The cows had given their milk; or rather let it be taken, so it was time for their hay. This tale would not be told to his peers at school. Why should anyone think he was afraid in the dark?

The light and dark side of kerosene lamps

After her mother died, Georgiana T. lived with her grandmother. She remembers they had to clean the lamp chimneys every night. Many of the persons interviewed old enough to remember using kerosene lamps agreed that cleaning the lamp chimneys and trimming the wick was a messy job. Blackened fingers were hard to get clean, much like the lingering brown stain gotten from hulling walnuts. Cleaning the lamp chimney was left up to women and girls for the most part. Some men remember having to lend a hand at it once in a while, but it was usually not for men and older boys.

Since Don D. had gotten up early before daylight to do the farm chores, he needed to carry a kerosene lantern to feed the livestock. In the feeding areas they had places to hang a lantern so they had both hands free for the feeding. At times the lantern had to be moved to bring light into the dark corners.

Dot H. remembered the oil lamps and then the wonderful change to a much brighter steadier light from Aladdin lamps. These lamps had a higher chimney and did not create the smoke of oil lamps. It was a plus for those who disliked the messy job of cleaning the lamps. However, Aladdin lamps needed to be handled with care because of a thin filament that was easily destroyed if touched.

Dorothy D. recalls some kerosene lamps were very decorative. With the transition to electricity some of them were converted to electric lamps. All the material needed could be purchased at the local hardware store. With the addition of a length of wire, a plug, switch and light bulb some families had a functional reminder of how it was back in the dimly lighted nights not really too long ago.

The bright side of a place in the dark

Shadows of the dark spots beyond the reach of kerosene lanterns and lamps were user friendly for children who, for various reasons chose not to be heard or seen. Wesley G. once garnered a treasured tidbit of information this way. At night children were supposed to be in bed, and not a party to adult conversation. Because of the dimness of the kerosene lamps, Wesley was able to slip into a dark corner of the room one night and listen. While secreted in a protective corner away from the light of the lamp, he was privy to the adult conversation of his grandfather and others seated in the lamp light of the same room. They were discussing with some disdain, the actions of certain young people of Wesley's generation who raced around in cars. Wesley listened with great satisfaction when his grandfather reminded those with him of the times when his generation raced with horse and buggy.

Ice was nice

On a hot summer day we heard a loud "boom."

All was quiet in our house. The TV quit. We did not hear the familiar hum of the refrigerator or our air conditioner. The microwave would not work. We had no power to run the washer or dryer. The laborsaving robots that we take for granted were strangely silent. We heard no happy hum from any of them. As we looked out our window to the street we saw some of our neighbors outside looking around. The power outage had come to their attention as well. What do we do now without electric power? We do what we seldom do anymore. Anne and I go out and have a good visit with our neighbors.

What's wrong with this picture? Why are we at home listening to the chatter of our refrigerator, washing machine, dryer, TV and computer rather than visiting with our neighbors twenty feet away? Why do I have the feeling they feel the same way about their family of electronic robots and laborsaving devices? What are we doing with the labor we have saved?

Being the senior citizen of our neighborhood, I feel obligated to tell them about the time before we had electricity. Not knowing how long the power will be off we all agree it is best not to open the door to the freezer or the refrigerator. That gave me an opening to talk about the "ice-box." Although there are those, with good reason, who think I don't know what I'm talking about, I've had the good fortune of talking with others my age in preparation of this book. They remember the ice-box.

Gene T. remembers his neighbors going down to the river to cut ice. River ice was cut into big squares weighing as much as 100 lbs then put in a small ice house. Sawdust was put under and all around the blocks of ice as insulation to keep them from thawing. Protected in this way the ice stayed solid until used in the summer. As to refrigeration in Kentucky, they didn't have an ice-box in the house. Their food for the winter was mostly canned or salt cured. Gene's family did not have an ice-box until they moved to Ohio.

Georgiana T. did not live on the farm, but she

agreed with Gene's account of the ice harvest. On cold winter days people cut ice from ponds in this area and also covered it with sawdust to be used in the summer. She marveled at the strength of ice men who carried 100 lbs. on their backs to be put in the ice-box in their house. Delivery men wore a leather or rubber apron on their backs to protect them from the cold water coming off the ice.

Tom W.'s little red wagon served for more than a recreational toy at a time when his family lived on the edge of Port William, Ohio. He has vivid memories of his role as a very young ice man. Tom's Mother Ora gave him a quarter to take his wagon to get a block of ice for their ice-box. First he had to go into Minnie's restaurant and give them the quarter for the ice. When it was paid for, a man took him to the ice house. This was all new for Tom. All that ice was still cold even though it was a warm summer day outside. The man explained to him about the wood chips on the ice that kept it from melting. No doubt others saw this proud young man pulling the load of melting ice on his way home. Pretty flowers, bugs, birds nor barking dogs slowed the young ice man on his homeward trek that day. The iceman delivered the ice.

Virginia B. recalls they had an ice-box before electricity. Perishable food was kept cool in the ice-box on hot summer days. After getting electricity and a refrigerator, did her family call it the refrigerator or the "fridge" as some do today? Many of us who went through the time of transition to electronic appliances did as Virginia's family. Their first refrigerator was called, like ours the "ice-box." Old habits like good friends stick with you.

Then there was light—and it was good

Before electricity came to rural areas some farm homes had carbide lights. Gene T. knows about carbide lights that burned with a cleaner flame than kerosene. In some places gas was piped into the house from a generating tank. Combustible carbide gas was generated from pellets dropped into water in a tank.

In the early 1900s the electrical energy business was booming in the cities. Soon the small towns had some of the same benefits. The rural areas however, did not get electricity until much later, after President Franklin D. Roosevelt created the Rural Electric Administration in 1935. Thanks to the genius of Charles F. Kettering some farm families had lights from a DELCO light plant. (Dayton Engineering Laboratories Company). The Delco was a 32 volt generator powered by a gasoline engine. The energy was stored in 16 large lead-acid batteries. When the batteries were fully charged, the plant automatically shut itself off. With the power drained from the storage batteries the engine could be self-started again to restore them by turning on the starter switch. In most cases the Delco did not provide enough power to do more than run small motors and electric lights. Dot H., Virginia B., Bob Mc., Jim H. and family, Susanne K., and others interviewed, grew up on farms that had a Delco prior to REA. Young people on family farms with a Delco light plant were often sent to "start the Delco." Since it had its own electric starter on the gasoline engine, all they had to do was flip the switch.

Burdette Q. reported it was a long time before they got electricity on some of the farms. Farm families had to deal with the argument of the power company's claim it was too expensive for only one farm. When they finally got six families to sign up, the power company began to put up the poles and lines. That was a memorable day! Light and power was soon to bring dramatic changes to the farm and home.

Tom W. did not have electricity on their farm until 1947. After they got electricity they kept kerosene lanterns in place for times when the electric lights went out due to power failure.

Electric lights and more power to you

The transformation to power was no less dramatic, but took a little longer to replace physical energy with the power of an electric motor. Those who lived on the family farms in the thirties and forties remember the difference made by: washing machines, irons, vacuum cleaners, cream separators, sewing machines, water pumps in house and barn, milking machines, motors to generate power around the barn, lights, baby chicken incubators and brooder heat, cooking and refrigeration.

Working in their hardware store, Mike D. learned the wisdom of keeping track of purchasing trends. He recalled when people finally got electricity on the family farm the first thing purchased was obviously light bulbs. The next most important purchase was a washing machine. The success of the business was in keeping stock of items in demand. New electric poles and lines in sight along rural roads in the 1930s and 40s were of great satisfaction for the local merchants in town as well as the families on the farm.

Turn your radio on

Many families had a battery operated radio before they had electricity. Families gathered around the radio after evening chores and supper to listen to favorite programs like "Amos and Andy," "Fibber McGee and Molly," "The Lone Ranger," "Little Orphan Annie," "One Man's Family" to name a few. Robert Mc. and his parents sat by their radio in 1929 for a very special program aired on WLW. All work ceased. Any interruption had to be of utmost importance to take them away from listening to a 15 minute weekly broadcast of Admiral Byrd from the South Pole.

Why do you think they called it a party line?

Many farm families had telephones before electricity. Press A. was engaged in private conversation on their party line phone when he heard a third party voice say, "Shut the door!" He knew his nosey neighbor was listening in on their conversation again. Her children in the next room apparently were making too much noise for her to hear all she wanted to hear. It was really "none of her business." Sometimes asking her to hang up because it was personal and private worked. Sometimes it didn't. You never knew for certain. That's the way it was with party lines in the rural areas. Anyone with a phone on the party line could pick up the phone and talk or listen at will. All on the party line knew this to be the case and accepted the fact that the phone system was used for eavesdropping and gossip as well as personal conversation. The party line phone was of utmost importance as a way to call the neighbors for help in a time of emergency.

Robert Mc's family was on a 12 party line. For most rural families "listening in" was an acceptable way to learn the community news. Much as e-mail today, having a number of people at a time "listening in" was an effective way of sharing important information of concern to all.

On God's time

One of Robert Mc.'s neighbors bought a clock on regular sun time and put it on his kitchen wall. When those in charge came out with Central Standard Time, he bought another clock set for standard time and put it beside the first clock. In the late 1920s time in this area was changed to Eastern Standard Time. That meant another clock to be put up in the kitchen. Then during WWII they decided to go on Daylight Savings Time, so he bought the fourth clock and lined it up beside the other three. Many farmers called D.S.T. "Damned Silly Time" and worked by the daylight of the sun. Before having electricity, farm families lived and worked by the light of the sun. They worked hard and were tired, so they did not stay up late at night.

Heart of The Home
Life and Equipment in The House

Women's labor centered on the house, men's work on the fields. The two met in the barnyard where divisions were less clear. ...A woman's responsibilities included maintaining the house, supplying food and clothing for the family, caring for the children, and supervising the work of those who helped her. ... Women tended to poultry flocks and oversaw egg production for household use or for trade in local markets. Women gardened, primarily to provide food for their families; sometimes, however, they specialized in vegetable or fruit production for locals markets.

Mary Neth—*Preserving the Family Farm*

In a time of so many broken promises and families we need to reexamine the heart of the rural family, the home nurtured within a house on the farm. Every family member in one way or another was a participating member. With a wide variety of labor opportunities each had a part to play in the health and happiness of the whole.

Monday's—Wash Day
Tuesday's—Ironing Day
Wednesday's—Mending Day
Thursday's—Cleaning Day
Friday's—Shopping Day
Saturday's—Cooking Day
Sunday's—The Lord's Day

For farmers big and small, breakfast had it all

If you lived on a family farm you probably had "dinner" at noon and "supper" at night. Urban families often had "lunch" at noon and "dinner" in the evening. While there was agreement on the subject of "breakfast" in the morning, breakfast was not to be taken lightly on the farm. It was often served in the "heart of the home," the kitchen.

Milk in a bottle fresh from the cow

Don G. has a heart warming family story of his early childhood. One morning when he was a toddler he awoke before his mother Edith had returned to the house from her regular chore of milking the cows. When he did not find his mother in the kitchen, he woke up his sister Wanda, 10 months younger, and started to the barn in search of their mother. Before leaving the house he picked up his milk bottle and gave another to his sister. Hand in hand carrying their bottles, the two toddled into the barn and found their mother still in the barn milking one of the cows. Since she had not finished her milking, she carefully squirted streams of milk from the cow's teat filling the children's bottles. Making them a little soft bed of straw, she watched her children nurse their bottles of warm fresh milk. Edith finished her morning ritual in front of the four approving eyes of her two contented children.

Breakfast on the go? A no, no—breakfast, then you go

For some it was grandmother, or mother, for others it was grandfather or father that was first to get up in the dark hours of early morning. Donald D., Gene T., Pat H., Margaret H., Wesley G. and Lois H. all agreed the call "Rise and shine!" meant everybody got up for breakfast. While father and older boys and at times hired men went out for morning chores, mother and older girls worked fixing a big breakfast. Donald D. remembers having pork chops or ham, eggs, biscuits and gravy or toast and coffee most every morning. Coffee beans were home ground. The grounds put in a cloth bag, were let steep in a big graniteware coffee pot. The pot filled with water was brought to boil on their wood burning cook stove. At times they also had cereal and fruit. On the farm you always had a big breakfast, unless you were sick.

Margaret H.'s mother, father and all connected with the family farm were blessed by a meal routine that made it easier for all to plan the day. With very few exceptions breakfast was served early in the morning, dinner at noon and activity on the farm stopped for the evening meal at 6:00 p.m. Once in a while during the busy seasons, Margaret's Mother Mabel took meals to the farmers in the field. As mother's helper to serve the farm workers, Margaret always enjoyed this time of a picnic in the field.

Pat H. has fond memories of breakfast at her Jones grandparent's farm home. It was a great way to start the day. When Grandpa Roy finished the morning milking, Grandma Ethel would have the bacon and eggs fried and ready for breakfast. The meal on the plate looked and smelled so good. Before they began eating, Grandpa Roy always read a chapter out of the Bible and finished with a prayer. Pat had trouble concentrating on what was read and said. Her mind was on the food waiting on her plate. It had prior claim to a healthy hungry growing girl. Pat was sure while grandpa was reading and praying, the bacon and eggs on her plate were getting cold. But that was not the case, as she remembers the food always stayed warm and delicious. It made a lasting impression on her. She was blessed with great food for both body and soul.

Meal makers

Wesley G.'s grandfather Taylor often wakened Wesley in what seemed to be the middle of the night. It was something of a mixed blessing. When you are having a good night's sleep it is not too pleasant to be jarred awake by any kind of noise. You could forgive the early morning shock of the rooster's crow. God created the rooster to do that. It was his thing. The banging of pots and pans was something else, until Wesley recognized it as work of his grandfather getting breakfast ready. He knew grandfather had dressed by the light of a kerosene lamp and was starting a fire in the kitchen woodstove. Grandfather was preparing another of his superb breakfasts. How could he go back to sleep when he knew breakfast would soon be filling the house with wonderful aromas?

Don G.'s grandmother Mary lived with them until her death when Don was nine years old. Don remembers his grandmother getting up early to start a fire in the kitchen woodstove before making breakfast. She did most of the cooking on a woodstove. Wood was kept in a box beside the stove. Grandmother asked Don's father Robert to get more wood when the wood box was empty and needed more kindling. He'd go out to the wood house and split off enough to fill the wood box by the stove. Filling the wood box and carrying out ashes from the cook stove was never ending. These little chores did not always come at the most convenient time. However, when you were hungry, you remembered it was an important part of the daily routine of the family farm. Splitting wood had the challenge of an art form for some and was never boring. For others it was a matter of doing what was needed to fill the box. Some people had more important things to do, until it was time to eat again.

Dotty R. and other friends have memories of the Daisy butter churn. One of the many activities often involved the children at an early age, was known as "Churning the butter." To be technically correct it should have been called, "Churning the cream." Not many people today give much thought to the miracle of milk. Most people who go to the dairy counter don't think too much about the product or the process. At one time milk was either whole or skimmed. Two percent or fat free was not given much consideration. Cream, real cream, rose to the top. If you did not care for cream on your cereal, cream could be saved for making butter. Milk left to sour (clabber), made different types of cheese. Cheese made at home was called "Cottage Cheese," because it was made in the home and not the barn I suppose. It is no longer made in homes much, so why is it still called "Cottage Cheese?" I've gotten off the subject of children and churning butter.

The first butter churn I used was a two-gallon crock with a round lid that had a hole in the center for a wooden plunger. The plunger was about 2 ½ feet long, round like a broom handle with wooden cross paddles at the bottom. Children were able to make butter by churning the cream lifting the plunger up and down. Usually it took a while with other children taking turns.

Dotty R. and others remember the Daisy churn as the first churn they used. It had two basic parts, a clear glass gallon wide mouth jar and a hand cranking mechanism with wooden paddles. If you kept turning the crank the paddles churned the rich cream into buttermilk and butter. When it became too hard for children to turn the hand crank an adult finished the job. Other butter churns were simply gallon jars or cans rocked back and forth by hand.

Jim H. remembers how hard it was for him to turn the Daisy butter churn hand crank when he could see the butter forming around the paddles. What started out as fun soon became boring. Churning butter was a chore children could do and be praised for the good work. Becky G. in the 40s used a Tupperware container that she rocked back and forth until she had butter. Her mother Mary thought watching too much TV was a waste. However, she allowed Becky to watch while she was making butter.

Something special for lunch

It took a little longer to have the butter ready for lunch than Robert's Mother Eva had expected.

Today many younger people have little idea of the process of making butter in the days before refrigeration. Even families who were not in the dairy business had a cow for milk and cream. For that matter people today know little about real cream. For the uninformed, cream is thicker, but lighter than the rest of the milk. Cream rises to the top of the can or crock. At this point it can be skimmed off and put into a butter churn. Given enough agitation in the churn, cream solidifies into butter. Warm milk from a cow must be cooled to keep it from spoiling. How do you keep milk cool without an ice-box or a refrigerator? Robert's family used the best cooling agent they had, cool water from a deep well. Robert had the job of pumping cool water into a low tank in the house basement that held crocks full of milk and cream. When the water began to loose its cool, he had to drain the tank and fill it with more cool water from the well. This was one of his regular chores as a young member of the farm team.

Eva chose to make butter before fixing their lunch. The churn she used was in the shape of a small wooden barrel. It was supported in a horizontal position by a pedestal in the middle and rocked by hand. Cream poured into the barrel churn sloshed back and forth as the ends alternated rocking up and down. When agitated in this manner cream eventually forms the semisolid mass of butter. At times it took longer for cream to turn to butter than others. This was one of those "take longer" days. Finally the slosh, slosh of the cream became more of a thump, thump with the teeter tottering of the churn. Eva had butter.

Because she was late for lunch, she did not put the butter into a mold as was her custom. This time she would dig a big spoonful out of the churn and take it to the table for their lunch. The butter was passed and spread on bread by each member of the family. No one, not even Robert's mother knew this butter was special. Unknown to all, a visitor had inspected the cream before it was put into the butter churn. When

Eva went down to get the rest of the butter, she was the first one to learn of the inspection and meet the inspector. If by chance, you are eating while reading this story, I suggest you finish eating before reading the "rest of the story." Let me put it this way: an overly curious and not too cautious mouse came to its demise falling into the cream and became coffined in the butter. It would have been interesting to be a little mouse in their house when Eva told her family about their very special lunch.

School lunch and a snack before evening chores

Like many others Dotty R. had the real KFC (Kitchen Fried Chicken). I'm not the one to tell another chicken fryer that "K." can mean a place closer to home. Her mother frequently had fried chicken on Sunday with enough left over to take to school the next day.

Family farm young people probably not unlike their city cousins, were hungry when they came home from school. Don G.'s grandmother made six loaves of bread every week. Don would take a knife and go to the smokehouse and cut off a slice of smoked ham. Then he'd go to the bread box and slice two pieces off a loaf of grandma Mary's bread. With the homemade bread covered by homemade butter and the ham in-between, Don had his favorite after-school snack. Now over 60 years later, he still remembers, "That was good eating."

David H. like Don his neighbor, also made an after-school snack. His was a treat of a generous amount of "sandwich spread" between two slices of bread. David still likes a sandwich made that way. After the snack he was always in a hurry to go out and see what happened around the farm. Chores always went better after a homemade snack.

The art of killing and dressing chickens

The first step in the preparation of a chicken dinner

is not for the faint of heart. Her parents died when she was young so Virginia B. and her brother Rendel lived with their Quaker grandparents. With a religious conviction that all life is sacred, taking a life was never easy, even if it was just a chicken for the noon meal. After Virginia was married she was on her own when it came to killing and dressing chickens. When a chicken is killed the goal is not to damage the body so the obvious way is to cut off the head. Some people took a hatchet or a corn knife to kill the chicken. With the head of the chicken on a chopping block one quick chop did the work. Others with strong hands or feet pulled the head off. Virginia's way was to hold the chicken on the ground. A sturdy stick was placed across the neck of the chicken so she could stand firmly on the stick. With her feet on the stick and the feet of the chicken in her hands, she pulled off the chicken's head. It was not a thing she liked to do, but if you're going to have a chicken dinner this is one way to begin.

One day she had three chickens to kill and dress. Virginia decided it best to go out of the yard and into the old orchard to kill the chickens. Out there she would not have the blood in the yard where one might walk. She carried the chickens with their legs tied so they'd not escape before the killing. In the orchard Virginia found a stick stout enough to hold the head secure. One at a time she untied the legs before removing the chicken's head. Each chicken was left to flop until the muscle contractions ceased.

She was picking up the last of the lifeless birds when the sight and sound of movement grabbed her attention. The sows! Virginia had forgotten the sows at the other end of the small old apple orchard. Free to run in the orchard, large mother hogs feasted on the dropped apples. That all changed when they got a whiff of chicken blood. Hogs will leisurely eat corn and apples but they came running at the smell of fresh meat. Virginia knew if she were to get out of there unharmed with bleeding chickens, she would have to grab them up quickly and run for the yard gate. Her speed was driven by the sound of their excitement. The woof, woof of hogs is something like the combination of a bark and deep growl of a large angry dog. Only a fool would try to drive off this blood thirsty pack. The problem for Virginia was the distance to the gate and the greater speed of the chasing sows. What to do? Realizing she could not out run them, Virginia used her head as well as her feet. She dropped one chicken for the sows and continued her race for the gate with the other two chickens held securely in her hand. The sows stopped where the chicken dropped. When she was safely in the yard beyond the fence, Virginia stopped to catch her breath. She turned to look back at the sows, and saw them fighting each other for a bite of fresh chicken. One chicken was soon turned into pork. Under the circumstance 2 out of 3 is not too bad. Sometimes, it pays to use your head.

CFC–Camp Fried Chicken

Usually, when a person puts sorghum molasses on biscuits it is not for fun. The satisfaction of pouring the molasses comes from the anticipation of the treat for your taste buds. Molasses made by Wesley's grandfather Taylor, could have been marketed under the label of "Good Time." Like butchering, processing sugar cane was a family and neighbors affair in the 1930s and 1940s. After cane stalks go through the crushing mill the sugar water is run into an evaporating pan. Under the large pan is a carefully attended wood fire. It takes a long time and continued watching to get the right heat for quality syrup. While the sugar water was evaporating the family and friends had their evening meal.

Wesley's grandfather had killed a chicken and removed the innards. The next thing he did came as quite a surprise to some who had not seen it done before. Taylor covered the chicken, feathers and all with clay from a nearby creek clay-bank before putting it to bake in the coals of the fire. After awhile he took it out and broke off the baked clay. The feathers and pin feather hairs were gone. Stripped of feathers and coat of clay was a beautifully baked bird with an aroma that said, "Come and get it." What a picnic! What a meal! The real fun lasted long into the night. While waiting and watching the evaporating, grandfather Banks played the guitar as people sang and visited. The long night ended up when "cane sugar water" had a name change to "Sorghum Molasses."

Farmers in the area also grew sugar cane and brought it to his grandfather's mill to have it processed. Tom W.'s father Tommy grew sorghum cane once as an alternative, when other crops were low in price. They cut the cane with a binder and took it down to the mill of Wesley's grandfather. Taylor Banks had a model A Ford. He removed a back wheel. In its place he put a pulley belted to a pulley on the mill that crushed the cane. The sweet juice of the cane was then boiled down to make sorghum molasses.

From hog to ham and in between

Hog butchering was not a "do it yourself" affair. It was one of those times of "neighbor helping neighbor," when Donald D. was a boy. He looked forward to these times of rural socializing and celebration. It was a fall celebration of the end of harvest and the satisfaction of working together. When the butchering on one family farm was complete, usually in a day, time was set for repeating the routine at the home of another.

The hog or hogs to be butchered were penned up the night before. At the break of day men and boys arrived with their own sharp knives, the main tools for the process. The hog was hung by the hind legs on a scaffold so the blood would drain from the slain animal. Near the scaffold was a vat of boiling water with a small fire that had been started earlier that morning. When the blood was drained, the carcass was lowered into the vat of scalding water. The butchered hog was left in the hot water for a few minutes. This made it easier for the men to scrape the hair off the hog hide. It was done on a temporary table made of long boards over a couple of sawhorses. Here the carcass was cut up into large pieces of meat, sides, shoulders, hams etc. The outer fat was trimmed, along with meat scraps, and was cooked in a large iron kettle heated underneath by a wood fire. When the fat had been heated to a liquid state it was put into a lard press. Donald described it as a large cylinder 16 inches in diameter. The solid top had a screw in the center. The screw was tightened to press the liquid fat out through holes in the bottom of the cylindrical press. The hot liquid was caught in a can with the end product being semi-solid lard when it cooled. Children standing by waited for the meat pieces from which the lard had been squeezed. It was a delicacy to be eaten on the spot.

At this point the women usually took over. Hams, shoulders and bacon all had brown sugar, salt and spices rubbed into the meat, with appropriate spices for each kind of meat. Meat to be cured was wrapped in paper or a cloth sack. Butchering was in the fall when the weather was cool. With help from family and friends much was accomplished in a day's time.

During the depression of 1928–29 hogs were selling for three cents a pound. Burdette Q.'s dad Frank loaded up 13 large hogs and took them to the slaughterhouse in Martinsville to be butchered for home use instead of selling them at a loss. They brought the meat home and put it in their smokehouse. It was quite a sight. Sugar cured hams and shoulders were piled high almost to the ceiling. Burdette's family had plenty of meat to eat for the winter. When they needed money for food not home-grown, some of the meat was sold to neighbors.

Mike D. described how sausage was preserved in crocks in the "spring-house." Sausage patties were stacked up in a crock. Cheese cloth was used to strain hot lard which was poured over the top of the meat. When the lard cooled it formed a seal over the meat. In the cool of the "spring-house" sausage kept without spoiling for three or four months.

Garden produce and food preservation

Dorothy D. remembers when she was young some of the garden's produce and fruit in season was picked, prepared and eaten the same day. No artificial preservatives were necessary. Many agreed with Dorothy, food prepared and eaten is this manner was far superior to what we have today. We can get almost any food we desire today 365 days a year. However, in most cases the flavor and quality does not measure up to the produce of the family farm. Farm families usually had big gardens and flowers in abundance.

David H. echoed Dorothy's testimonial concerning fruits and vegetables. They were at their best in season. He believes we are spoiled today when we can have most anything we want in or out of season. With fruit and vegetables that are picked and shipped before they are ripe, we do not get the same taste and quality. We've become impatient victims of "we must have it now," and have forgotten the satisfaction of waiting for the special treat of the seasons best.

During the depression years of the 1930s Donald D. remembers farmers and their families got along better than some because they had food, water and fuel without needing much money to pay for it. With little financial output they survived by having large gardens and methods of food preservation for the months when there was no garden or orchard produce.

Bob Mc.'s Aunt Laura dried sweet corn using an enclosed metal container. It was square, about 3 or 4 inches deep, with a place to be filled with water and heated on the cook stove. Corn was cut off the cob and put on top of the heated container. The heat of the water dried the corn. Dried corn was then put in a bag to be used with water added later as needed. Other fruits and nuts were dried and preserved the same way such as persimmons, papaws, hazel nuts, walnuts, hickory nuts and chestnuts. Vegetables, potatoes, onions, etc. and apples were kept in dry cool partially underground cellars.

Dot H., Press A., Dotty R. and Mary Lib S. spoke

of preserving food by canning. This was one of the most important ways of being assured of food to eat in the long months of no garden growth. Most remember the rich variety of color in clear glass jars of produce on the shelves in a pantry or cellar. Chicken, pork and beef in jars joined the vegetables, jams, jellies and fruits. Canning was a hot late summer job usually done by the women and girls of the farm. A walk into the pantry or cool cellar affords a feast for the eyes. What a rainbow of colors from the deep red/purple of canned beets to the bright yellow of freestone peaches! The beauty for the eyes is excelled only by the taste to the tongue. Those who have spent long hours in the garden and a hot kitchen can truly appreciate the personal investment of love and hours of hand labor from the topsoil to the tabletop. The labor and heated work of canning was relieved to some degree by the introduction of the community frozen food lockers in conjunction with places of livestock slaughter and meat processing.

Dotty M.'s father Isaac always carried a gun when he went out in the field to kill rabbits for the family to eat. Many depended on adding variety to the menu with rabbit, squirrel, pheasant or quail.

Clothes of the times

Press A. told of the importance of overalls for farmers. They came with practical extras. Bib overalls had pockets for a pencil, knife, pocket watch, a bandana as well as a loop to hold a hammer and pockets for gloves and tools. A youngster felt he had arrived when given his first pair of overalls. I've never understood that "pair" bit. It's only one isn't it? The same goes for pants, but not for a shirt. I wonder why? I've gotten off the subject again. Farm girls wore dresses, except farm girls who worked at the barn or in the fields. Girls working on the farm also found Bib overalls to be practical and comfortable. This was acceptable working attire on the farm, but not to be worn to school, church or town on Saturday night.

"Homemade," was not a matter of being fashionable in the 1930s. It was a practical answer to the shortage of money. Georgiana T. recalled the popular practice of making clothing and other household items from printed feed sacks. During the time when cotton prices hit a new low, innovative merchants started putting their company names and logos on cotton sacks with pretty colorful designs. That which began as advertising products such as flour, sugar, rice and beans became material for women to use in making clothes, bibs, diapers, handkerchiefs, dish towels and other items for household use. It had special appeal to women at a time when making do was a necessity. Farm feeds and seeds were also sold in sturdy cotton sacks with colorful prints suited for many household needs.

Today products for human use or consumption often have a "shelf life" printed on the container.

During the 1930s and 40s clothing for growing children had an unwritten "self life" clause in the "hand-me-down clothes" regulations of clothing ownership. A new garment for the first child was his or hers only as long as it fit. Then it was passed on to the next child in the age chain. Burdette Q. recalls the time during the depression of the 1930s when younger children did not get new clothes as often as their older siblings. In those days when money was in short supply, "hand-me-down" clothes with holes or worn places like elbows and knees were not justification for new garments. Holes were patched and knee and elbow pads of sturdy material were sown to those places subject to more wear and tear. Virginia B. tells of the alteration of dresses and coats cut to fit the smaller children. Parents sensitive to this invitation to sibling rivalry, made it up to the younger children in other ways in keeping with their needs and desires.

Tom W. remembers his family did not have many clothes in the 1930s. His everyday attire was usually bib overalls. Houses did not have many clothes closets in those days. Few rural people had a large collection of clothes. Clothes were winter or summer, work or Sunday (good) clothes. Smaller quantity meant fewer clothes to wash.

Early Monday morning—washing clothes

Most of those interviewed had running water in the 1930s and 1940s. A few remember the labor intensive times of washing clothes without running water or washing machines. However, they do remember as late as the early 40s women tried to get the wash out on the clothesline on Monday morning. Monday was wash day.

If washing clothes is not one of your favorite things, you need to consider what it was like doing the wash before electricity and running water. As a general rule men and boys didn't do the washing. Some may have helped by filling the iron kettle in the backyard with rain water from a barrel used to catch the water of the roof drains. The water was heated by a hot wood fire under the kettle. While the water was coming to a boil, clothes were sorted in three piles, whites, colored, work clothes and rags. A cake of lye soap was shaved with a knife and dropped into the boiling water. To give the clothes body and ease of ironing starch was used. Starch was made by stirring flour in cool water and thinned with boiling water. White things with dirty spots were soaked in warm water and scrubbed on the scrub board. Colored things were put in warm water and rubbed before putting them in the rinse water and starched. A broom stick handle was used to take clothes out of the boiling water. Clothes out of the hot water were put in the first tub of cold water. Those fortunate enough to have a hand cranked wringer found it much easier than wringing water out of the clothes by hand. The end result was a line or two of colored and white clothes waving in the breeze for a quick dry. Rinse water was put on the flower bed. Boiling water was used to mop the porch. Turn the empty tubs upside down and washing was done. Washing was done, but there was the little matter of ironing the starched clothes. That could be done tomorrow. Remember Tuesday was the day to iron clothes.

Donald D.'s family had a washing machine that was a laborsaving step up from the cast iron tub in the back yard. Theirs was a single tub with a hand cranked central agitator. Donald's dad rigged it up so it would be even more laborsaving. He hooked up a small gasoline engine with a belt attached to a pulley in place of the hand crank. They still had to use the hand driven clothes wringer until he was able to figure out how to hook it up to the gasoline engine also. It was very crude, but it did the job and was much better than the old washboard. Donald knew many farmers who were innovative in their creation of laborsaving devices.

Dorothy D. noted that after washing clothes, most farm families hung clothes to dry on the clothes line in the backyard. In some cases clothes were hung there in the winter as well as other seasons. Freeze dried clothes could be kept outside until brought in a little at a time and hung in the kitchen by the heating stove to finish drying.

Saturday night bath—need it or not

Hands and faces were washed daily using a cloth and water in a small basin. Feet were washed as needed. Complete baths were a Saturday night affair. The Saturday night bath prior to hot running water was something else. Like most wood fired cook stoves, Donald G.'s family kitchen cook stove had a water reservoir on the side. The kitchen cook stove needed to be wood fired every day to have water heated for washing dishes and baths. Most of the children took baths in a large tub situated in the kitchen near the water reservoir of the cook stove.

In the heat of summer baths for David H. and his family were at times taken out under the stars and in the cover of darkness. A tub placed out in the grass was filled with warm water. All three children took their turn in the same bath water. Not to be wasted, it was then used to water the flowers.

Keeping warm in the winter—hot it was not

A time for unseen socks

It was not easy for Wilma Q. to get to sleep that winter evening. Nothing seemed to be right in her world. Before a school day, children had to go to bed when it was dark outside. The darkness came much too early. Wilma was not ready to go to bed. It was cold in her room. How could she sleep on those cold sheets? However, most of her worries were about tomorrow. Her mother Emma had helped put out her school clothes and had included those hideous long brown stockings! The next morning her mother, in spite of Wilma's protest, insisted she was to wear the stockings to school because it was winter and it was cold!

When the bus came Wilma walked out into the cold and got on the bus wearing those ugly, long brown stockings. Once in the bus out of sight of the watchful eyes of her mother, she had it her way. If she took them off, someone would be certain to catch her in the act and word would get back to her mother. Wilma decided not to take off her stockings. She had another plan. When others were looking away, she gradually rolled each sock down to the top of her shoes. It worked for her. No one at school took notice of her ugly socks. On the return trip home she pulled them back up. Wilma walked into the house with the socks in place like they were when she left. Her mother did not discover her secret. Friends, I won't tell if you don't.

Snow removal in comfort

Jim H.'s snow shoveling experience began when

he was quite young. The house where he lived as a child was not completely weather proof. Fresh air was in good supply on a cold, windy winter day. When it was snowing outside the house, it was snowing inside as well. It was in the kitchen that Jim got the idea of shoveling the snow that had blown in around the loose fitting windows. Jim was happy and quite content to shovel snow with his small toy shovel and dump truck. It is probably safe to say, the end result was more water mopped up than snow removed. The essence of the day's activities was a happy youngster having fun. Years later Jim would labor with deeper snow and a larger shovel. Is he still happy about the snow? You will have to ask Jim.

Some like it hot—some like it cold

"Warm Morning" was the name of the coal/wood burning stove in the living room of the farm house of Sharon G.'s youth. To be truthful the stove needed to have "Cold Nights" added to the name. We cannot blame the stove. It might have been warm at night as well, if someone carried in coal and wood at night as was done during the day. That would not have been practical at a time when all able family members had responsibilities during the day and needed their sleep. Like it or not, the cold facts remained. It was cold especially in the two bedrooms upstairs. Cold feet were given some degree of comfort with heated flat irons wrapped in heavy towels when put at the bottom of the bed.

Fine feathered friends

One of the blessings of rural living in the 1930s and 40s was people did not have to stay up for the 11:00 p.m. nightly news. For one thing, today most of the news is anything but comforting and conducive to sleep. The old saying, "Going to bed with the chickens," was a reference to the time, not the place of retiring. Sleeping in the barn was a possibility for me, but don't send me to sleep in the chicken house!

(You probably think I have gotten way off the subject again, but I haven't). Gene T. in a way slept with the ducks and the chickens. I'll have to explain this right off, or I will be in trouble with Gene.

Gene's family raised ducks and chickens for food, but that's not all. The down of the ducks was put in their pillows and the feathers of the chickens were used for feather beds. Like Gene, I also slept on a feather bed. I rest my case. Did we not go to bed with the chickens? Many others in the 1930s slept under or over feather beds. They helped on those cold nights when no one was minding the "Warm Morning."

Late to bed—early to rise

I am not certain this story of Tom W.'s belongs here or in the section on recreation with pranks and jokes. The story had a cold point as you will see, so I put it here with the cold nights.

Being a teenager, Tom was late getting home one night. He did not go into detail telling why he was late, but Tom was late enough to feel the need of getting to bed without waking others of his family. He did not know at the time, there was one member of the family still awake and for good reason. Tom's sister thought she had good reason to be awake. Tom may have thought otherwise, but I'm getting ahead of myself telling the story.

It was all dark when he entered the house. That suited Tom as he sneaked around being careful not to bump into furniture getting to his room. He took off his clothes and put on his pajamas in the dark. Then Tom quietly got ready to slip into bed. When he crawled in and began to pull up the covers trying to get comfortable, Tom warmed up to the fact that his bed was lumpy and cold. It was terribly cold! With no consideration given to keeping quiet, he jerked out of bed and turned on the light. What was that cold form under the sheet? With the light on, Tom pulled back the bed sheet. The ice cold lump he was lying on was a big icicle his sister must have pulled off the house and placed in his bed. Why did he think it was his sister? Why did he hear her giggling in the next room? Tom didn't tell me the rest of the story and I didn't ask.

In the dark of the night

It had been a long day working on their farm for John H. He had no major problems, just a day of minor interruptions. After supper he reclined in his favorite easy chair to read the latest issue of his farm magazine. It didn't take long for his eyes to close, his head dropped as if trying to get a closer look at the magazine now resting on his lap. At his wife Ethel's insistence he finally went to bed. It wasn't long until he was captive in the land of deep sleep. When Ethel joined him an hour later, she was comforted to see the regular rise and fall of his chest accompanied by snickering snore with every intake of breath.

In the time of deep sleep, John was strangely aware of a degree of discomfort. His thinking was encased in a fog that entertained the struggle between, "something is not quite right" and "I'm just too tired to deal with it now." Was it a bad dream? John was not sure where he was or what he was doing. However, this would not last long. The cover of the fog was about to be blown. But it would not come from John's effort to rise above it. There was another who would bring John back to the real world in time and place.

David E. was a good friend and frequent visitor of John and Ethel's son Noel. He helped them on the farm and stayed many nights with them. It was not unusual for David and Noel to be out and about after supper. This evening the boys came to bed well past the time of usual retirement for the family. No lights were turned on when David slipped into the bed saved for him. He thought about stopping in the bathroom first, but to do so would mean going by John and Ethel's bedroom. David did not want to wake them so he crawled into bed hoping to sleep.

After a time of off-and-on sleep, David realized he had to go to the bathroom. Carefully walking down the hallway trying his best to avoid the squeak of the floor boards of an old house, he made it safely past John and Ethel's bedroom. He had no plans to turn on any lights that might disturb others sleep. It was dark in the bathroom, but he decided not to turn on the light in there as well.

David backed in as he pulled the door shut and started to sit down on the stool. In an instant David stood up. The seat was much too soft and not the firm surface he'd expected. In the darkness it all came to light for both David and John. All it took was for John to stand up where he had evidently fallen asleep on the stool and ask the obvious question, "Dave, what are you doing in here?" What did David say? What could he say but, "Nothing yet John."

None shall make them afraid

Gene T. told us in Kentucky they grew broom corn and made their own brooms. People swept their yards with big brooms to keep the snakes away that might be hiding in the grass. Some people may have lived in fear of snakes. Others knew they were out there and did not want them in the house, but they didn't lie awake at night in fear of them.

The same goes for people in the community. In most every community there were a few unsavory characters, but not many people lived in fear behind locked doors. Few people on family farms bothered to keep their doors locked night or day. Even when farm families left for a few days their doors were not locked. Neighbors came over to feed the livestock and check on other things. Farm neighbors looked out for each other. Family farm homes had a well worn path which usually led to the back door. The back door was often the door into the kitchen. Friends and neighbors entered by the customary path unless it was a special occasion. That's how it was in, "The Heart of the Home" for farm families.

Breeds and Brands
Working With a Choice of Livestock and Equipment

Countryman's God
Who reaps the grain and plows the sod must feel a kinship with his God:
For there's so much on earth to see that marks the hand of Deity.
When blossom springs from tiny shoot:
Where orchard yields its luscious fruit:
When sap is running from great trees on all occasions such as these
The man who breathes fresh country air must know full well that God is there.
Roger Winship Stuart—*Masterpieces of Religious Verse*

Work with horses, hogs, cattle and poultry was an essential part of growing up on a diversified family farm. Children had ample opportunity to observe and understand the cycle of life. They knew the source of their food and the sacrifice in labor and life from which it came. They understood how their life also was rooted in the soil. Children had opportunity to develop affection for animal pets and accept responsibility for their care. Adults had a choice of animal breeds and machinery brands. At times farmers had to weigh their preferences against what was financially sound and best for the family.

Committed to caring for livestock

Once in a while on the farm we were confronted by situations of challenge to our civil vocabulary. After it was all over, more often than not, we had a funny story to tell.

Cows in small herds on family farms usually had names. For the most part they were common names, like Buttercup, Daisy, Lady, or names of women such as Betty, Maud or Elsie the Borden Milk cow. If the cow came from registered stock she would have a name from her family line. When Ray C. was asked, "Did you give each of your cows a name?" He said, "No we did not have registered cows so they had a chain around their necks with a number." Ray went on to add a second thought saying, "however, at times I did give them a name I'd better not put in print."

Cattle come and go

Margaret H. lives on the farm of her youth. One of the farm features still in good repair is the old barn. Still standing beside the old barn is a silo that brings back memories of the livestock of days gone by. Livestock on their farm was primarily chickens, pigs and cattle. The cattle came in to Port William, Ohio on the train. Margaret's Dad Alton went to Kansas once to buy the cattle. The cattle that he purchased were shipped by train to Port William. After that Alton had someone he knew in Kansas buy the cattle for him and have them shipped. When the cattle arrived, Alton would go to Port William and with his hired help take the cattle off the railcar and walk them 2-3 miles to his farm. The cattle were fed grain, hay and silage in the barn lot. When they reached market weight they took another ride to market. This made room for the next group to be fattened on their family farm. The silo that was used to store corn silage for the cows now stands as an empty monument to a time when cattle feeding was a profitable option for family farms in this area.

What makes a fair, fair?

It is obviously impossible to give every boy and girl equal opportunity in competition for the ribbons and prize money at the fairs. Children with caring parents, have a better chance than those who do not. Conversely, children with little interest or effort with parents determined their offspring win top honors do not deserve winning rewards. When Dick G. was in 4-H effort was made to give each child equal opportunity at least at the start. In the fall the leaders of their 4-H Steer Club would call a meeting of everyone interested in a Beef project. Usually there were about 125 boys whose parents helped them buy a steer. Those in charge and others, who could afford to go, went to Montana and bought all 125 steers at one ranch. They rented a railcar and shipped the steers back to Ohio. When they were ready for distribution, a number was put on the tail head of each steer. Each boy chose a steer by drawing a number. This arrangement was not based on how much money a child or his family had. At this point at least, the child whose father was a successful owner of a fine herd had no advantage over those who did not.

All the steers were the same breed for every participant. The club alternated every year between Herefords and Angus. Dick's dad Clarence had Angus cattle. The years the cattle chosen were Herefords, Dick and his brother were supposed to keep them in the barn. Clarence didn't want anybody to know they kept anything but Angus on their farm.

In our interviews we had general agreement concerning the importance of equal opportunity for children showing the fruits of their labors at the fairs. It is not right for one child who happens to have parents with money and high quality livestock to get all the honors when so many other children worked just as hard without parental help and received little financial reward.

Anne E. told of a time when there was no preferential treatment based on parental influence. Her father was recognized as one of the successful farmers of the area when she showed a "so-so" Hereford beef calf at the fair. Anne's calf placed 24th out of 25 entries. The calves of the County Agricultural Agent's son were placed 23 and 25.

David H. works full time off the farm. He has a herd of over forty beef cattle and a flock of sheep. He raises the calves and feeds them out to market weight. The sheep and cattle are the means of keeping the grass and weeds under control as well as producing some meat, mutton and wool. When his children were younger the animals were their 4-H projects and shown at the county fair. The children benefited from the responsibility of ownership and satisfaction of relating to the animals. He rents his tillable ground to the care of another. David misses the pleasure of actually working the soil and tending the crops, but the limited acreage of ownership makes it economically impractical. The expense of owning and maintaining the equipment for tilling the soil calls for contracting the work to another.

Little boy blue, come blow your horn

David H.'s comment: "If you have livestock long enough, they will break out of their confinement." Usually it is at a most inopportune time. Most every beef or dairy farmer with a few years experience of ownership and responsibility for cattle has heard the unwelcome words, "Your cows are out!" Under the best of circumstances, some cows will find a way to break out. When this happens, usually the whole herd thinks it's a good idea and follows the leader. With the help of family or neighboring friends, it is only a matter of a few hours until the fence is fixed and the bovines without borders are bound where they belong.

When David got the call on Sunday evening telling him thirteen of his cows were out and on the move, he didn't think it would take long to get them back where they belong. Little did he expect a week later he'd still be waiting for his cows to come home! Thirteen just might be an unlucky number. That evening they were safely confined in the pasture of a neighbor. The plans were to corral them the next day and truck them back home. This was not to be. When David and brother Jim tried to load the cows the next day, one of them was probably spoofed after being shocked by the pasture's electric fence. The cow managed to tear the fence down and take off. The rest decided to do the same. All thirteen cows raced into an unfenced wood without as much as a goodbye wave of their tails.

In the week that followed the front page story on the lips of the neighborhood was, "Have you heard about David's cows?" Numerous sightings proved to be after the fact, or of neighborhood cows that were neither lost or out of place. It is one thing to get a sighting on runaway cows, but quite another to get them to stand at attention. Curious animals will run from a stranger, but then amble back to get your name, address and reason for the visit. Not so with these Clinton County cows on the run. The cows visited neighboring corn and bean fields, but spent little time in open fields that week. Neighbors came to join the search. Even the Boy Scouts offered their help.

Fortunately for the cows hoping to stay hidden, corn was at its best. The corn stalks were tall with leaves filling in the space between the rows. People walked along the corn rows and heard, and at times saw the cows. But how do you corner them in a cornfield? The noise and sight of strangers added to their fear and cause for flight. Those in the search found out how hard it is to catch cows running between the rows of corn. Frightened cows do not even respond to their master's voice. One of David's cousins tried to locate the cows by flying his airplane over the cornfield where cows were sighted. Flying just above stall speed did not afford much of a view of the cows. No doubt the sound of a low flying plane added greatly to the cow's discomfort. Some were so distraught they might have given serious consideration to joining the cow of nursery rhyme fame in a jump over the moon.

Do You See What I See?

This would require waiting until the dark of night and they did not have that much time, so the cows continued their cornfield run.

Farm families with cattle have an appreciation for the old saying, "waiting for the cows to come home." It is never easy to put aside the worry that one or more of the cows trying to cross a busy highway, might cause a serious accident. After you have done your best and failed to capture the cows, you leave it up to the cows to decide they'd had enough. That's what David, family and friends did. With all the patience they could muster, they waited.

David purposely left the gate open to the field where the cows belonged, hoping they would come looking for the security of the home-place. The wayward cows proved it was worth the wait. Whether cows were divinely led or just came back to the comforts of home is a mute point. The cows came home. A granddaughter of David's brother Jim saw them first as they passed the family farm. She said to her grandmother Lois, "Look, there are Uncle David's cows!" Lois stopped the car immediately and called David to report the good news. "Your cows are home!" David was on his way home from work in Xenia, Ohio. Driving carefully but something above the speed limit, David thought to himself, "If a patrolman signals for me to stop, he will have to follow me all the way home. My first priority is to get that gate closed as soon as possible." He was not stopped on the way. With an energized shove, David closed the gate on the wayward cows. Grateful to God and helping neighbors, David had the best night's sleep he'd had for many nights. Relaxed in rest, he knew what he had to do the first of next week.

Monday morning out of consideration for his family, friends and neighbors, David treated the cattle that toured the neighborhood to another outing. This time they rode in style in a truck to the stockyards of Hillsboro, Ohio. The sheep are still in the meadow, but the cows are not in the corn.

When animals get your attention

Jim, David and Becky H. grew up on a truly diversified livestock farm. On their farm they had horses, beef and dairy cattle, hogs, chickens, ducks, sheep and guineas. Bob, their father milked 20 or more cows. He sold grade B milk in 10 gallon cans which were kept in a milk-cooling tank, until picked up by the hauler each day. The tank was also a good place to keep watermelons cool. Thanks to the tank, what could be more heart healthy than cool watermelon on a hot summer day.

She was not exactly sheep-shape

Jim H. told at times they put calves in with the ewes in the barn at night and in the day as well during the winter months. One heifer calf hung out with the sheep. She seemed to think she was just a big sheep, even though she was actually a small cow. As she grew older, she was confused and even went into bellowing fits. Jim feels she had identity problems not being certain whether she belonged to the sheep or the cows. I've had times like that; haven't you?

Just try to catch me!

When it came time to capture sheep to trim their feet or worm them, Jim could usually get the sheep to come into a pen for a little grain put in a pan on the ground. One older sheep was called "Wilder," for good reasons. She was wild and refused to be trapped into the pen. She was a smart one who sensed this was not your ordinary free morning meal. Wilder knew Jim was up to something. She'd stamp her feet in defiance. The "catch the ewe" game was on. In the pasture where he kept the sheep were a lot of bushes and she could use them to avoid capture. "Run around the bush and catch me if you can," was the theme of the day. The challenge for Jim was to outrun the ewe.

In the pasture field there was a wagon, and the sheep to be caught knew she'd be hard to catch if she got under or around the wagon. Jim caught the look in her eye as the ewe took off. He knew he'd have to outrun her to the wagon, or she'd have the better advantage. Wilder beat him to the wagon. She thought she was safe under the wagon until Jim decided to roll under to drive her out. The wild ewe stayed safely just out his reach. He slid out from under the wagon and tried to drive her out throwing sticks and stones. When the ewe finally slipped out, Jim knew he'd have to keep running until he caught her. Eventually Jim's persistence paid off and he was finally able to grab her. After a few years Wilder settled down and became the "wise one" and grew to appreciate all Jim did for her. She seemed to realize the game was over.

Where sheep may safely graze

Sharon G. grew up on a diversified family farm. She was the youngest of seven children and had few required farm chores. Never-the-less, she liked to hang around the barn. Sharon fed the sheep and helped milk the cows. They had a mixed herd of 12 Holsteins and Guernsey cows. They also had sheep, 100 ewes, as well as chickens and turkeys.

Maternal deception

With sheep at times a ewe gave birth to only one lamb, and the lamb died at birth. At this time of lambing, other ewes had twins and triplets. In an attempt to match the ewe that lost a lamb to one of the twin lambs not getting enough milk, Sharon's mother Annabelle would cut a swatch of wool off the dead lamb. She would sew some cloth on the wool to be tied on the undernourished twin. This lamb was then taken to the ewe that lost her lamb. Sharon and her mother always hoped the ewe would accept the lamb that had something of her lifeless one on its back. Sometimes it worked and sometimes not, but it was always worth a try.

The trick for ticks takes sheep by surprise

Sharon G. told about fighting ticks on sheep. Sheep

ticks were a problem for the sheep in the area where her family farmed. After the sheep were sheared, they were corralled. At one end of the corral they had a ramp that opened to a tank below. The tank was filled with water and treatment to kill ticks on contact. The sheep had no problem running up the ramp to freedom, but they were not prepared for the wet bath detour in route. Sharon's family kept all their livestock in the barn during the winter with a fenced-in lot to the outside. The doors were closed at night and opened during the day if the winter weather was not too severe.

Way ahead with a better back

Jim, David and Becky's dad Bob H. started shearing sheep with hand shears when he was 13 years old. He had a strong back and was flexible enough to bend down and touch his elbows on the ground. Bob could stay in that position for a long time if need be. Jim helped by catching the sheep and shearing their belly. His dad then did the harder part shearing the rest of the sheep. Bob could shear 14 sheep an hour. Jim, without any help, sheared his neighbor's flock of 35 sheep and it took him all day. He was really hurting at the end of the day. Jim could not bend over like his dad. He often got down on his knees to finish the job. Bob sheared sheep for his neighbors as well. One day he sheared a total of 98 sheep for three neighbors. That included the time it took to move to each neighbor. At the end of the day he had to come home and milk his cows. While relaxing at milking he thought, "Why didn't I go out and shear two more of our sheep so I'd have a total of 100?" Bob was not as fast as some shearing sheep, but he was able to shear more in a day's time because he stayed at it. While others took a break to stretch an aching back between sheep, Bob stayed with it two or three hours without much of a break.

Bob started with hand shears. Later he used shears powered by an electric motor with the shears on the end of a shaft. When the hand held electric shears with electric cord attached came to be the latest innovation for sheep shearing, he used them some, but preferred the shears on the shaft. For one thing with the newer method, the sheep kept getting tangled up on the electric cord.

After the sheep was caught, the art of sheep shearing began by setting a sheep on its rear with its back to the shearer. A person's knees were used to hold the sheep while shearing the belly. A good shearer had a pattern that was repeated for every sheep to be shorn. A good job resulted in all the wool coming off in one piece. Bob was a good shearer. He did a good job. The most telling thing his family and neighbors knew, Bob was a good man who knew the "Good Shepherd."

What do you want with your coffee?

Lambs are sometimes weak at birth and need help to make it with a stimulant. The "old timers" were known to bring a weak lamb around with a little whiskey. Some "teetotalers" who would never let the stuff touch their own lips might shudder at the thought of getting a little lamb hooked on alcohol. Press A. has good news for those who seek a better way to give a weak lamb a warm welcome to a long life. Give them a little bit of coffee. Press not only claims he was told by a veterinarian, but he used it for weak lambs and it worked for him. The claim is coffee is an even stronger stimulant than whisky. With the help of a little coffee, weak lambs will get up on their little wobbly legs and soon be hopping around full of energy. I'm old with wobbly legs. Perhaps I should drink more coffee.

The pride, praise, profit and pain of pork production

The white out

The care and feeding of 1,700 hogs is a major operation when they are not confined to one building with equipment for self-feeding and watering. Even if he had the space and equipment, Burdette Q. likes to see his hogs enjoy the freedom of open space.

It was mid-winter in the 1950s when the area was hit overnight by a fairly heavy snow fall. Burdette had not expected to see the snow on the ground this late in the winter. He got dressed in his winter clothes and prepared to check on the hogs. As he walked through the fresh white snow to the lot where he expected to see the hogs, he was brought up short by what he saw and what he did not see. All Burdette saw was a level field of white. What he didn't see brought a lump in his throat and a rapid beating of his heart. Burdette did not see his hogs, not one! What happened to the hogs? There was no way some person, or persons could have loaded up and sneaked off with that many hogs without him hearing them. Hogs let the whole world know about it if they are being forced to incarceration in a truck. As much as Burdette hated to face it, the only other explanation was the hogs had gotten out someway. Even with good fencing it was always possible for a hog to find or make a hole in the fence and invite the rest to join in the escape.

What to do? It was foolish to try since the hogs obviously were out of the range of his voice, but out of habit he called them. "Whooeee—whooeee, piggy, piggy, piggy." In an instant the white ground in front of him began to shake and 1,700 hogs stood up from under the cover of their blanket of fresh white snow.

Pampered pig—pet or pest

"You don't want that pig. With that broken leg, he won't amount to much." Gene T. was about 10 years old when he heard a neighbor had a pig with a broken hind leg. He asked his neighbor if he would sell the pig. "I'd planned to kill it," he told Gene. "But if you really want it, I'll give it to you." The pig and its problem now belonged to Gene. Guided by imagination and powered by ingenuity he took pop-sickle sticks and made a splint. With the pop-sickle splint in place around the break of the leg, Gene wrapped it in place with tape. The pig lived. The leg healed and a boy had a pet. The grateful pig followed Gene like a faithful dog everywhere he went. To hear the pig tell it, he and Gene were inseparable buddies. The pig had no problem joining Gene in the house, perhaps with the hope of sharing a meal at their table. Gene's mother Julia was not a party to the presumptions of the pig. Gene and the pig had a couple of "run-ins" that with the guidance of a broom in the skillful hands of Grandma Julia, became "run-outs."

No welcome mat at the house of this hog

Inspectors seem to have a habit of appearing at inconvenient times. For some farmers there was never a convenient time for an inspection. The same was true of some farm animals. We had Jersey bulls. We never expected to be welcome to an inspection of the bull barn while the bull was home. The bull made it clear to us he was in no mood for an inspection.

Tom W. found this to be true at the home of a sow with a litter of baby pigs. His dad pulled a hog box in the middle of a pasture field as the home for a sow about to give birth. The sow gave birth to a litter of ten baby pigs. She was not in the mood for a visiting inspector of her home or offspring.

One day guided by his curiosity and without his parent's knowledge, young Tom took a leisurely stroll out to the hog house to pay a visit to mother and babies. He meant no harm. Tom just wanted to have a look at the little litter. Something about his unannounced visit did not set well with the sow.

At the sight of this intruding inspector the sow jumped up and out of the hog house. With an open mouth the sow showed what big teeth she had. With a warning "Woof," she prompted Tom to give his visit a second thought. He was quick to get the rebuff of her second "woof," and saw no reason to let the sorry old sow know why he called. If you'd been there, you would have understood his decision to turn and head for home as fast as his little legs would take him. Unfortunately four legs are faster than two. The sow's take on the story was, "the boy tripped and fell down right in front of me. I couldn't turn fast enough to miss running over him." The truth as Tom as tells it, was the sow knocked him down and ran over him. I have to believe Tom's version since the sow died a few years later and was not interviewed by a reporter at the time of the accident.

Obviously, Tom came through the ordeal with only a few scratches. He did sustain a scar and scare, both of which have remained with him to this day. Tom is not a hog farmer. He has no desire to even the score and no desire to have hogs on his farm. What about a hearty helping of ham? Now that's a different matter.

It was not good when push came to shove

If you don't understand what it means to "hog-down" your food, there is no better way to understand it than carrying a five gallon bucket of buttermilk to feed a group of hungry hogs. Heaven help you if it is your first time to be in an open lot trying to pour buttermilk into an open trough in the company of a host of hungry 300 pound sows. A dozen or so sows all pushing and shoving while you are pouring buttermilk into the trough, is a lesson in poor manners you will never forget. There is no place in the book of swine survival for "You first." Donald G. found this basic "Me first" drive in hogs to have tragic consequences. Donald did not want his hogs to be raised in confinement. During good weather the hogs enjoyed the freedom to range and eat out of the confinement of pens, but they did have a place in the hog barn to get in out of a storm. They had the swine equivalent of the popular "doggy door" for canine pets. The hogs had the choice of going in the barn or out to a fenced-in lot. Don was away from the farm when a storm approached one afternoon. His hogs hurried for shelter in the barn gaining entrance through the hog sized door. It is not clear how the door was shut, perhaps by a gust of wind of the storm, or the rush of the hogs

Do You See What I See?

in an attempt to escape the onslaught of wind and rain. But the door was closed and thirty market sized hogs were suffocated or trampled to death by the frenzied charge to escape the fury of the storm. The same basic drive for feeding survival prompts panic unto death. Unlike hungry hogs, let us be thankful for the spirit that guides us to courage and consideration of others.

Yes you can and I can too

Rosalee W. called "Tudy" by her father, enjoyed being outside on the farm with her Father Eddie, known as "Warpy." She even went with him when he was working with the livestock. Her father had hogs and liked caring for them. Eddie even made pets out of some of them. But when baby pigs were with their mother sow, they were not to be petted. Mother sows are funny that way. Rosalee had no fear when she was with her father out among the hogs even though some almost matched her in height. She felt proud and protected. Her pride and bravery were challenged one day as they walked up to a hog box in the open lot. It was the home for one of the pet sows. The sow about ready to give birth to baby pigs did not come out of the hog box as usual to greet them. Her father wondered if the sow already had her pigs. When he heard a defensive "woof" from the sow, Warpy sensed the danger to his young daughter. He shouted, "Hurry Tudy climb up on the hog box!" Rosalee replied, "Daddy, I can't, I just can't," but when the mother sow came charging out of the hog box, Rosalee made it. To this day she is not certain how panic pushed her to the top out of harm's way. Rosalee's panic turned to curiosity while she watched her father look in on the sow now back with her pigs. When her father got a close-up look at the menacing teeth of the still excited sow, pet or no pet, he knew what he must do. Warpy jumped up and joined his daughter on the top of the hog box. Together they watched the old sow cool down and go back inside satisfied she had saved her young. When they were certain the sow had forgotten their uninvited visit, Rosalee and her father slid off the roof and hand in hand walked slowly to the safety of their home. Can you imagine what her mother Grace said when Rosalee told her of their afternoon adventure? You will just have to guess. Rosalee did not tell me.

Poultry in place on the family farm

The chicken train

I've heard of "The Milk Train." I have had no personal encounter with one. It's my understanding that the name was given to slow early morning trains that made many stops along a route. If I am wrong, I'm sure some reader will set me straight. When Bob Mc. told me about his uncle and the "Chicken Train," that was a new one for me. Bob's Uncle George worked for a man in Blanchester who bought and sold chickens, eggs and cream. The market for chickens was not as great in this area as other parts of the country. Most families around here had their own chickens and had little reason for buying a live chicken. I assume the owner of the business had heard of a market for live chickens in the cities. That being the case, New York City must need a lot of chickens. The B&O Railroad tracks in Blanchester in one way or another ran east to New York. A railroad car loaded with live chickens just might bring a lot of money if they made it all the way to New York. Live chickens placed in wire pens were not too happy about the trip. Blanchester suited them and they had no desire to travel to New York City. It was a long way to New York even by train in those days. Unhappy chickens would be in a "fowl" mood if they did not get food and water all along the weary way.

Here is where Bob's Uncle George came into the picture. A Chicken Train needed some one to go along to see that the chickens were fed and watered so they would be delivered alive in New York. Another little matter needed to be taken into consideration. A railroad car load of unattended chickens might arrive at its destination, without the chickens. Uncle George was the young man chosen for this enchanting job. He rode in a small space on the "Chicken Train" as nurse maid to a huge flock of chickens being taken for a ride. When arriving at the end of his New York trip the only way back to Blanchester was to sell the chickens until the car was empty. Like everyone with a regular job, when George arrived back in Blanchester, he had another ticket to travel. A railroad car was ready to be loaded for another ride on the "Chicken Train" bound for New York.

You turkey!

If a person wanted to be really insulting one can always bad mouth the turkey. To be called a turkey is no credit to the bird. Sharon G. knows from personal experience that turkeys may be some of the biggest birds around but by far not the brightest. According to her family's agreement with them, turkeys were supposed to stay back in the field where they were housed. Their turkeys did not see it that way. They kept coming back around the farmstead and proved to be a nuisance around the house and barn. What can you do? You can't drive Turkeys.

One day Sharon watched some turkeys run to examine a walnut that fell near them. The birds didn't try to eat it. They were just curious. A plan for moving the turkeys back where they belonged took shape in her mind. If she threw a walnut ahead of the turkeys in the direction of their home perhaps they would go after it. Aha! It worked. With one walnut after another thrown ahead of them the turkeys went back where they belonged. This worked for a while until they decided to come back and it would have to be done all over again. Were the turkeys dumb, or did they find it amusing to see her picking up all those walnuts? I'm glad Sharon stayed with it. It's a "win-win" situation. Exercise and turkey meat are both good for a healthy heart.

Ways of watering for every season

Did you ever wonder why Benjamin Franklin or Thomas Edison did not get into a pack of trouble by the curiosity that prompted them to try so many dangerous experiments? My curiosity got me in trouble time and time again when I was younger. It began to slow down in my 70s when I ceased to be able to follow curiosity's directions. I understand Rick K.'s agony when he was swimming in trouble following the call of his curiosity. That's the way it is. Curiosity lures you into trouble, but is nowhere to be seen when you need justification for your actions.

I'll "woof" and "woof" if you tear my house down

Water for the livestock was pumped by a windmill. Water pumped by the windmill went into one large tank and then into three other tanks. The system was surrounded by four fenced lots with a watering tank in each lot. The flow of water from the main tank was controlled by a float. As a young person in his early teens, Rick K. was fascinated by the system and played in the water tank. The trickery of trouble slipped in while he was cooling off wading in the main supply tank. All was well with the water level in each tank. Curiosity urged Rick to take a closer look at the float system to see how it worked. Curiosity took a leave of absence when for some reason the float stuck. Water kept running and would soon be running over where it was not needed or wanted. Try the best he could, Rick could not get the thing to quit. He felt he had no option but to hurry and get the help of his father Alfred. Rick jumped into the farm truck nearby, got it started and headed for the barn lot. Little did he know more trouble lurked ahead. In his haste, plus lack of truck driving experience, Rick let the truck drift into newly plowed soft ground. When he tried to get it out with a sharp turn of the steering wheel, the truck whipped around striking a hog box housing a sow with little pigs. The sow came out woofing and the pigs squealing, as their home collapsed. If there is a good ending for the story, it did not all happen that day. His father fixed the float, but it took some time to repair the hog box. The little pigs soon got over their scare. Rick's dad did not send him away to work for another farmer. However, it was some time before the mother sow had anything good to say about Rick's truck driving skills.

Pumping water with windmills

On Donald D.'s family farm water was pumped into a stock watering tank by a windmill. In the winter they would chop a hole in the ice for the livestock to drink.

Margaret H.'s family also had a windmill that pumped water for the livestock.

Wesley G. remembers on his grandfather's farm water for the livestock came from a nearby spring. Even in the winter, his grandfather carried all the water for the livestock from the spring in a bucket by hand.

Creek water works well until winter

A creek ran between the house and the barn on Dick G.'s farm. Water for the livestock came from the creek year around. The animals were free to drink from the creek at will except when ice froze in the creek. The creek was fed by a spring. In winter their dad Clarence had the boys carry water in five gallon buckets for the livestock in the barn. When the boys began each took one five gallon bucket. Clarence would not allow them to carry only one five gallon bucket full at a time. That would be a waste of time. Their father reminded Dick and his brother Don they both had two hands. They could carry a five gallon bucket in each hand. At the time Dick and Don felt this was cruel punishment. After they were older they saw it dad's way. They realized it was easier to carry a balanced load in each hand.

On warming water for hogs

Under the heading of "Those Trying Times," Press A. recalls the challenge of raising hogs in the winter. His hogs had the shelter of unheated hog boxes. The hogs did well when they had a place to get in out of the wind and extreme cold. Feed was put in self-feeders with heavy metal lids over each individual feeding trough. The hogs soon learned to lift up the lids with their nose and let it come down with a "bang" when they were through. This was drum beat music to a hog farmer's ears when he was resting in his bed at night. The hogs were eating and that was good. It was not the same for their city friends when they came to visit. The banging of the hog feeder lids way into the night was not conducive to a good nights rest in the quiet of the country.

Having water for the hogs in the winter was another thing. The watering tanks, usually round, also had individual watering places with metal lids. Unlike dry feed, water freezes in a tank without heat. This problem was taken care of by a wick-lighted small kerosene lamp designed for hog water tanks. The lamps kept the water from freezing. Lighting them was no big deal, right? Think about it. Heat rises so the kerosene heaters did no good burning on the top of the water tank. The makers of these tanks thought about it and provided a little space below the tank to slide the lamp in for heating the water. No big deal? In the summer on a dry day when someone has cleaned up the hog manure around the water tank it's OK, but what hog drinks hot water with his meal? On cold frozen ground, or cold manure covered wet ground with a strong north wind blowing, it is a big deal! It is true a person does not have to stand on his head to do the job. All you have to do can be broken down into 4 easy steps.

Step 1—lie down in the cold muck with the mixed smell of hog manure and kerosene.

Step 2—lying flat on your back with a wet glove

removed from your hand, try digging the matches out of one of your rear pockets.

Step 3—find a match and try to light it. (A match can usually be struck by raking it across one of the metal overall buttons if you can reach one).

Step 4—reach in with a flickering match and persuade the fickle wick to take the light before the wind blows it out. If the match goes out and will not stay lit, repeat step 3 over and over again, until you have a lighted lamp.

Did I mention the friendly 300 lb. hogs nudging your knees and chewing at your boots? Love those hogs; it's just one of those fun things to do on a cold winter day. Ask Press A. to tell you about it. He's been there and done that. If you have a son who will own the farm when you retire, see if you can get him out of school to help as part of his winter vocational education.

The warm water wagon is here

If you didn't know better, your first impulse would be to call the fire department and tell them Burdette Q's water tank truck was on fire. On second thought you'd ask, "How does one have fire on a tank of water?" Never-the-less, the tank truck was slowly burning and also slowly moving down the lane between the pasture lots of Burdette's hogs.

Burdette had a novel way for watering hogs in freezing weather. On each side of the 1,500 gallon water tank were two tubes which he stuffed with burlap feed sacks. The sacks had been soaked with gasoline and set on fire to warm the water in the big tank. Burdette drove the big old tanker truck in the lane between the lots filling each of the smaller 80 gallon hog watering tanks. There is no reason to believe warm water makes for better pork, but without it hogs don't do very well eating snow and chewing on ice. The burning water wagon may have been puzzling to the "passer-by," but it was a welcome wagon for the hogs.

Harvesting corn—a different kind of "bank board"

I'm not sure a preacher or a teacher will back me up, but I know I can hear better when I close my eyes during a sermon or a lecture. All visual is shut out and the sense of hearing takes over. If I put my hands over my ears, do I see more clearly? I'm not certain and don't want to go there in public anyway. If I close my eyes and try to shut out the activity around me, it does enhance the eyes and ears of my memory.

Using these helps and the gift of memory I go back to the late 1930s. On a cool, crisp fall morning with the sunlight reflection of a multitude of tiny frost crystals, my sister and I stand waiting for the school bus. We hear the bus in a distance as it rolls over the complaining stones of gravel at rest on our county road. Another noise has greater claim of my attention and arouses my excitement. The regular repetitions of dull distant "bangs" were the sounds made by my father and near neighbors harvesting corn. How I wished it was not a school day. Perhaps Saturday, I could be out with him helping strip the shucks from the ears of corn. Once in a while I was able to hit the "bang-board" with an ear. Years later I learned it was called a "bank-board." The "bank-board" was a 2 or 3 foot high addition on one side of the wagon to keep ears of corn being thrown over the wagon.

Donald D., David H., Dick G., Tom W. and others recalled the satisfaction of harvesting corn by hand. Stripping the shucks from the ear was helped by a hand glove with a hook or a peg. The hook was inside the leather glove and the peg protruded upward between the thumb and index finger. Both were effective. It was a matter of personal choice. A team of horses pulled the box wagon with "bank-board" over the two rows of harvested corn along side of the row to be harvested. Those who were experts at shucking corn could rip enough shucks off the golden ear of corn to break it free from the stalk. Even before a toss of the shuck-free ear to the "bank-board," they were in the process of grabbing for another ear. Most farmers tried to finish harvesting by Thanksgiving, but wet weather usually kept them from husking corn every work day. Often they had to shuck corn in the cold of winter.

When Dick G. was old enough, he helped his dad shuck corn. Even when he went to college, his dad saved 10 acres for Dick and his brother Don to shuck by hand when he came home for Thanksgiving.

What will it be, "shucks" or "shocks?"

My dictionary in one place defines "shucks" as a mild expression of disgust. "Shocks" is the plural of a response to a major emotional disturbance. Use of these expressions today has nothing to do with corn harvest. The only connection I find to harvesting corn is that corn shucks are a very mild disruption to the stalk losing a yellow, shuck-free ear to the wagon. On the other hand, corn shocks are a clean separation of the whole corn stalk, ear, leaves and husks from the supporting soil. This little dictionary detour may not have made your day. It is simply a way of dignifying the terms that represent two ways, of harvesting corn, "shuck" or "shock."

We did not shock corn on our farm in Illinois, nor did any other farmers in our area. It was all new to me. I learned the "why and how" of it all from Tom W., David H. , Donald D. and others.

Tom W. told me the advantage of shocked corn over husking corn in the cold of winter was in placing the corn shocks somewhere around the barn where it was protected from the cold winter winds when it came time to shuck the corn from the shock.

David H. said shocked corn was cut when the ears were mature but still green leaves so it would hold together in the shock. It was not harvested as early as corn is today. Ear corn was left in the field in the shock until late in the fall to dry before being shucked and brought into the barn for feed and fodder. Some men went from farm-to-farm just to shock corn.

Those who shocked the corn usually had a rope used to pull the stalks together to make the shock. With twine they would tie the stalks together in the shock and release the rope to go make another shock. It was a matter of pride and money to see who could get the most shocks in a day.

Donald D. noted they used to cut corn by hand with a corn knife. He had a cousin with a knife attached to his shoe and cut the corn stalk by kicking it. Later they used a McCormick Deering binder to cut and tie the corn bundles. Gathering chains would bring up the cut corn to a platform where the bundles of corn were tied with binder twine. The bundle of corn was then dropped to the ground in a row near where it was to be gathered to be part of a shock. To give it an early start, wheat was sown on the ground around the shocked corn.

On a cold sunny day in January, Dick G. and some of his high school buddies worked shucking corn from shocks. As he recalls, time passed quickly because one of the boys brought a transistor radio. They worked out of the wind with the shock of corn as a windbreak and enjoyed the popular tunes of their day.

"A century ago, when horses hauled in the hand-picked corn crop, the rise of the harvest moon marked the end of the day. A good man and his team of horses could expect to husk and haul 50 bushels of corn to the crib in a grueling workday. Both horses and husker needed to hit the hay early to rest up for tomorrow's demands. Fast forward a hundred years, and the changes are staggering. Machines have conquered much of the physical challenge—a modern 12-row combine gathers approximately 50 bushels of corn each minute that it rumbles down the row."

Dean Houghton—*Moonlighters*—*The Furrow*

Don't give a hand to the shredder

Donald G., David H. and Tom W. reminded us that many farmers lost a hand or an arm and some their life in an encounter with a corn shredder. When a corn stalk got stuck in the rollers it seemed much easier to give the stalk a little push with your hand while the rollers were still turning. It didn't seem to take much to get the bunched stalks to go on through. Far too many farmers found once the rollers got a hold on the stalk or stalks held in their hand, a person rarely was quick enough to let go. It was at times fatal. At other times this mistake resulted in a loss of the fingers or the whole hand. Most of the time, these accidents happened when a person was in a hurry.

Don's father, Robert used a pitchfork once to try to push the stalled stalks on through while the rollers were turning and the tractor was on idle. The rollers grabbed the fork tongs and slammed the fork handle smacking his father in the mouth. Don's father was fortunate to learn a lesson instead of losing his life.

Women working outside the home

Dick and Sharon G. think capitalism with all its benefits has contributed to the decline of community and the family farm with its emphasis on making money. After World War II, many farm women went to work off the farm to help meet expenses for the purchase of improved laborsaving devices on the farm and in the home.

June C. worked as office secretary at a grain-storage and feed mill business. Wilma Q. worked in a doctor's office. Sue H. drove a school bus. Karen L. lived and worked on a pheasant farm gathering eggs and helped in other ways. She also helped in the hayfield loading the bales on the hay wagon. This was hot tiring work, but she enjoyed being outside on the farm. Margaret H. and Eleanor G. taught school.

What's your problem?

Mary Lib S. worked in a County Agricultural Extension office. In this position she obviously had a lot of contact with farmers and farming. The artificial insemination service for dairy cattle was in the early stages of its development. Mary Lib had to take calls from farmers who were embarrassed to talk with a woman about the matter. When they'd call many farmers would say something like, "uh, uh …I have a cow uh," Mary Lib would break in and ask, "When did you first notice she was in heat?" Then they felt a little more comfortable talking about it.

Anne E. had a similar experience when they had dairy cattle. Semen delivered by mail had to be accepted and signed for. It could not be left in the mail box. The mail man delivering semen for their dairy cattle would not leave it with her. He'd come back when a man was around if he had to, but he was too embarrassed to deliver the package to a woman.

My wife Anne's father had a fine herd of Holstein dairy cows. My father had a herd of Jerseys. Like many people who've had dairy cattle, the breed of choice is the one from your farm experience. As a rule, Holsteins give much more milk than Jerseys, but the butterfat, (cream) content per gallon of milk is less than Jerseys. We both know which is best, but it does not merit a heated debate.

Mary Lib S. had Jerseys on their farm. Once when she was visiting a large Jersey farm in Wisconsin, she noticed they had one Holstein cow in the herd. Mary Lib asked why they had the Holstein. The man replied, "We save her milk to wash out the milk buckets." For some reason Anne does not think this is true.

When the time is right for fishing

Don G.'s great uncle Truman was a farmer who loved hunting and fishing as much as farming, or maybe a little more. In his older years Truman had a small farm. One spring he was planting corn with horses, when some of his friends came by. All they said was, "Let's go fishing." Truman said, "OK," and tied up his horses and away they went. It's all a matter of knowing what's really important.

Saturday Night—For What It's Worth
Farm Related Businesses—Produce Sales, Grocery Purchases and Visiting in The Neighboring Town

Throughout the Midwest, farm people's purchases reflected a desire to increase their ability to communicate with others. Of the items counted in the census, farm people first bought telephones and automobiles. Both shortened the distances that separated farm people from neighbors, towns, and markets.
Mary Neth—*Preserving the Family Farm*

Town and neighborhood businesses existing for and by the support of family farms

RFD came before FDR

Both the coming of the Rural Free Delivery of mail and the "New Deal" of Franklin Delano Roosevelt were controversial. While the family farmers were very much in favor of mail delivered to their rural homes, they were not that supportive of the "New Deal" of Roosevelt. However, the small town merchants saw the RFD as agents of support of their business competitors the mail-order houses.

Prior to the RFD farm families had to travel to the post office in the nearest town over rural roads often muddy and rutted. The difficulty of the trips and uncertainty of having any mail made their visits infrequent. Farmer's organizations such as the National Grange were in touch with congress to approve free mail delivery to rural areas giving them the same benefits enjoyed by people in urban areas. With political maneuvering farmers were encouraged to petition their congressmen. Congress was overwhelmed by petitions.

RFD was proposed as an experiment in 1896 to provide rural areas the same postal benefits enjoyed by city and small town residents years earlier. Congress had approved the idea of a rural experiment in 1894, but opposition to the idea on the part of the Post Master General put it on hold. It was not until a new Post Master General William L. Wilson decided to try the experiment in his home state of West Virginia that Rural Free Delivery of the mail got its start. From 5 routes out of Charleston in 1896 RFD expanded into 43,718 rural routes serving 6,875,321 rural homes in 1930.

The RFD postal service with the help of rural people worked to improve the roads. Farm families began to see the benefits of better roads and appreciated the opportunity to have their share of road taxes reduced in exchange for their material and labor.

As feared by the local town merchants, farm families were sent catalogues giving them opportunities to purchase a wide variety of items at lower cost by mail. However, the road improvements necessary for mail delivery made it much easier for the farm families to come to town and see the actual item to be purchased. Town and country were brought closer together and farm families had greater access to the local, national and world news through the newspapers and other mail delivered to their home each day.

The mail-order catalogues were both a blessing and a challenge to the rural farm family. It was financially stressful to be constantly reminded page after page of the many things that would be nice to have. The challenge was strong for many young people to leave the farm for more money and a comfortable life. With the catalogue they could actually see the latest fashions and laborsaving devices. Mother's butter and egg money could not cover the costs of the urban lifestyle. Blessed were the farm families that met the challenge with a bond of relationship stronger than the whims of the fashion trend setters.

As the farm family became more dependent on money and the cost of making improvements on the farm and in the home, they had more interaction with the neighboring town. Saturday night at the end of a busy week became a time to go to town. Here they would sell or trade the produce of the farm for items of need not produced on the farm. For the children as well as adults it was also a time of visiting and enjoying the company of their town and country neighbors.

After the depression of the 1930s the self-sufficiency imposed by very limited income began to fade as an increase of family income made it possible to go back to town on Saturday night. Here farm families purchased items of need as well as some hoped for improvements. During the week and on Saturday evening farm related businesses such as grocery, clothing and hardware stores were open to meet the needs of the farm home.

Farm implement dealers, feed mills, grain elevators and blacksmith shops were in business to meet the needs of the farm. Livestock yards, meat lockers, ice storage and automotive sales and service were open for business in most towns. Add to that, banks, doctor's offices, barber shops, drugs stores, churches and a funeral home and you have some idea of businesses that depended upon the support of the farm people who came in into town on Saturday night. It was a two-way street. Merchants were in business to sell what farm families as well as families in town wanted and needed to buy. Merchants who succeeded recognized the need for socializing by providing facilities for visitation and promoting activities of entertainment.

Good to see you—how've you been?

Dot H. and others interviewed knew from personal experience going to town on Saturday night was part of a rural ritual. Bob Mc. worked in a grocery store. On Saturday nights farm families would come into the store to put in an order and leave for a while to go to other stores or visit with friends and neighbors. Before going home, the family would be back to pick up their groceries. Obviously there were no credit cards in those days, but people had credit. Many charged their purchases including groceries and paid the balance at the end of the month. No interest was charged. People were trusted except for a few who tried to take advantage. Word soon got out. The offending party either made arrangements for paying or lost their credit privileges with other merchants.

When Eleanor G. was a teenager and their family lived on the farm they always went to Greenfield, Ohio on Saturday night. Eleanor visited with young people her age. In town they often found a place to go and dance. At other times they simply walked around town to see others and to be seen.

Robert Mc. and others talked about Square Dances sponsored by the Grange that were a bit lively at times. Dances were held in a large room upstairs above Dunn's Grocery Store in New Vienna, Ohio on Saturday night. It was clear to the customers in the store below that something was going on when, according to Robert, grocery items started falling off shaking shelves. He also remembers the band concerts in the summer time in New Vienna and the free movies on Saturday night provided by the town merchants.

As Robert Mc. and his father Arthur entered the town barbershop, the long row of colorful cups caught his attention.

"Why does the barber have all those cups up there?" Robert asked his father.

"They are shaving mugs." replied his father.

"Why does he have so many?" Not waiting for an answer, Robert continued, "What does the barber do with all his mugs?"

"They don't belong to the barber. They belong to the men in town and out in the country. They come in here to have the barber shave off their whiskers. Each man has his own mug."

"Will I have a mug someday?" asked Robert. "You may, but not for a while. You will probably want to shave yourself. But now you are here for a haircut." answered his father as the barber motioned to Robert saying. "Climb up here young man," and "How old are you?" he asked.

Do you know what we did?

Mike D.'s grandpa Estel was not happy with his Dayton job. Talking over the matter with his brother who owned a hardware store in West Union, Ohio, he said "You ought to go into the hardware business. You won't make a lot of money, but it is a good way to raise a family." When Mike's grandpa heard about a store for sale in Ripley, Ohio, Grandpa Estel and Mike's Grandma Ethel drove south to Ripley to check it out. They were met at the door by a lady in charge of the store. After a warm welcome and introductions they told her why they'd come to Ripley. The lady was very helpful and showed them the store as well as living quarters on the floor above. It seemed to be perfect; just what they were looking for. While in the upstairs room they noticed a door which the lady said was a stairway to the attic. "Can we go up and look around," Estel asked. The lady took them upstairs to see the attic. Grandpa Estel noticed a lot of barrels and boxes which he discovered on closer examination were empty.

"What are these barrels for," he asked?

"We put the merchandise in them to keep it dry when the Ohio River floods. They float around until the water goes down," the lady replied. When she talked about the river flooding, that dampened their interest in a hardware business in Ripley. Nothing was settled on the sale of the store at the time.

On the way back to Dayton, they talked about what might be involved if they decided to buy the business. As they drove Ohio Rt. 62 out of Ripley to Rt. 73 it was near supper time. When they reached New Vienna, Ohio on Rt. 73, they saw a small diner. Tired and hungry Mike's grandparents stopped the car, got out and went in. A young girl, whom they later learned was the daughter of the diner owner, took their order and soon had it served. It was quiet, comfortable and they enjoyed the meal until a woman barged in as if she owned the place. It turned out she did. Mike's grandparents were in no way prepared for the turn of events that followed when a chain-smoking, foul-mouthed, nosey woman approached them. Was this the owner? Her first words were, "Who the H… are you, where are you from, and what the H… are you doing in New Vienna?" Taken back a bit, Mike's grandpa told her they had been to Ripley to consider buying a hardware store. Her reply was, "Why the H… did you go all over looking for a hardware store when there is a perfectly good one across the street for sale?"

They hardly knew what to say when the diner owner went to the phone and called the hardware store owner who sent his son to show the store to Mike's grandpa and grandma. They liked what they saw and the deal was settled that night with a handshake. The only stipulation was that they continue to employ Fred, one of the long-time workers. As it turned out, that was one of the smartest things that Mike's grandpa did. Fred was able to give grandpa, Estel good advice. For instance, at the close of WW II when women had not been able to purchase much in the line of kitchen cookware, Fred suggested they stock up a good line of enamel wear. The timing was right and the hardware business flourished. Grandpa Estel's hardware store was a favorite stopping place for residents of the town and surrounding farm community. On the trip home to Dayton, this unscheduled stop at a small diner had major consequences for Mike's grandpa, farm families as well as others living in the area.

Country doctors—I'll be right there

In the 1930s and 1940s, 911 was only a number after 910 and before 912, or in some cases a street address. In a medical emergency at any time night or day, you called or sent for the family doctor. If it required immediate attention, doctors made house calls. It was not uncommon for doctors like Dr. Buchanan to travel out to a family farm home even in adverse weather. Bobby and Wilma Q. were young parents. Their baby daughter had the symptoms of "croup" constant coughing and labored breathing. A snow storm drifted roads and made it impossible for them to get to town and see their doctor. With the help of a snow plow ahead of him, Dr. Buchanan drove out to the farm bringing comfort and care to the little girl and her anxious parents.

When he was in the third grade Robert Mc. was sick a lot. Doctor Matthews, their family doctor, came out every night when Robert was sick. On one occasion he looked at Robert for three minutes and then Doctor Matthews sat down with Robert's dad Arthur and talked politics for an hour and a half. Beyond their modest service charges, country doctors were repaid by the satisfaction of seeing friends return to health and having patients as personal friends.

Where do you go to rest and wait when you have finished shopping on Saturday night; you are ready to go home, but your children are not? A neighbor family of Margaret H. had five young daughters. The girls went with their mother to town every Saturday night. As their mother did the trading and shopping, the girls visited with their friends. There was little question about who was ready to go home first. The mother remembering the days of her youth let the girls have more time with their friends. Because of the thoughtful kindness of her family doctor she was welcome to wait for her children in the comfort of the doctor's waiting room. This courtesy may have been the idea of his wife. Country doctor's wives often shared their husband's concern for the health and happiness of their larger family of friends and neighbors. The girls' mother also had an opportunity to visit in the waiting room. Other mothers and some fathers shared the same time of waiting for children who never seemed to have enough time with their friends on a typical Saturday night in town.

What's up?—Saturday night recreation and socializing in town

In the small town in Illinois near where I grew up,

as in most small towns, high school age children ran around together when turned loose on Saturday night. On one such evening a group of 20 or more decided to form a "chain gang." Hand-in-hand the line followed in and out most anywhere the leader and his cohorts decided to wander. Most of the merchants enjoyed or at least tolerated the teen antics as the gang of 20 moved in and out of some of the establishments. At one point in the time of their winding wander they came up to the main intersection of the town. A car was stopped waiting to cross the main street. The car driver and his wife were local friends and known by some of the children. Since the car was stopped and there was no one in the back seat of the four door sedan, the leader of the gang opened the near door and crawled across the seat to the far door. After opening it the leader continued to drag the chain gang of 20 behind him. The last member of the giggling gang thanked the couple and closed the car doors behind him. Fortunately the neighbors enjoyed the prank as much as the children. The other cars on-coming and behind took no offense.

Unfortunately we are in such a hurry today; these antics would be no laughing matter.

Sharon G. told of a prank when she was in school. A group of boys put a herd of sheep in the school and left them over the weekend. (It's always the boys. Sharon denied any involvement). No major damage was done. It came as no surprise on next school day when the students were given the fun of the cleanup.

Dick G. recalled the actions of a small gang of big boys in town near their farm. The boys got inside a large open bottom chicken coup about 5 feet high, 4 feet wide and 6 feet long and walked it to the middle of a road in town. At that point in the adventure, the boys ran into a problem. One of them happened to notice the town constable walking up to the obvious out of place coup. The boys were caught in the coup and did not have time to get out. Quiet as roosting chickens they waited to see the constable's next move. While he was looking at it he called the sheriff on his two-way radio describing the chicken coup in detail. To the wonder of the wide-eyed boys the constable failed to look inside. When he sauntered away the boys slipped out undetected. An account of this wild adventure was to be told only to those who promised not to tell. Some of them had second thoughts and decided it was too good not to tell. You know how that goes.

Making it with milk and egg money

On many family farms extra money was made from the sale of milk, cream, butter and eggs as well as garden produce in season. This money kept separate from the income of the farm and fields, was the fruit of family labor often under mother's supervision. The money was hers to spend on food, clothing and household furnishings not produced or fabricated on the farm. Saturday evening was the time to go to town to buy, sell and socialize.

Money canned for safekeeping

Part of the challenge to careful use of the milk and egg money was impulse buying. In the stores on display for all to see, were the many "It would be nice to have," extras. What mother saw as only "wants" were translated to "I need one of those," by other members of her family. During the tough years of the depression of the 1930s there was very little money for "wants." Food on the table was mostly home-grown except for the staples like rice, flour, salt, sugar and spice. New clothes were out of the question when old could be patched. To help meet mounting expenses, produce of the farm was marketed in town.

It was not usually money in the bank. Often under mother's keeping, the temptation was to plead to mother for a share of the family treasure. Some fathers attempted to bypass financial requests by directing children to, "Go ask your mother." Sometimes it became necessary for many mothers to have a secret hiding place to make certain she had funds to meet the basic needs of the family. Georgiana T. lived with her grandmother. Grandmother had her cash stash sequestered in a Calumet Baking Powder can.

The can was somewhere in the smoke-house hidden among the processed meat. Georgiana did not say how she knew about this well kept secret.

Let them eat cake

With the culinary skills and artistry of a budding professional, Dorothy M. helped her mother Helen bake cakes for families in town. On the farm they had milk, butter and eggs. Using some of these basics with a few other ingredients, mother and daughter made cakes. Thanks to their home cooking skills the cakes were both eye pleasing and taste tempting. Dorothy and her brother rode with their mother in their model T Ford taking their tasty treats to town. The margin of profit over selling milk, eggs and cream assured Dorothy of another trip to town as soon as the cows and hens came across with more milk and eggs.

Up, Up and away—but not for long

A home basement is not usually considered a butcher shop, but it was for Bob Mc.'s mother Marie. Live, market weight turkeys, were brought to her by the grocer, a few at a time, to be dressed for sale. The live turkeys were kept in the security of a crate in a corner of the basement prior to preparation. After the turkeys were killed they were scalded, to make it easier to pluck the feathers. When the innards were removed, the dressed birds were cooled in tubs of water. Bob's father Arthur took the turkeys to be sold where he worked at McDermott's Grocery.

Marie and others who were helping her didn't know how the turkey got loose from the holding crate. It may have sneaked out as another wing flopping fowl was pulled from the crate. One of the turkeys decided this was not in its best interest to stand around and watch the proceedings. With a short quick turkey trot and a burst of wing energy, the turkey flew out of an open basement window and over the road to the sanctuary of the nearby cemetery. The turkey apparently decided to give its escape his best shot and landed up in a tree. Unfortunately for the turkey, his best shot was not good enough. A neighbor with a rifle actually had the last shot. The high flying bird was brought down to earth to join the other turkeys for someone's Thanksgiving dinner.

Extra milk—Mother Mabel's money

Margaret H.'s family did not have many milk cows. The few they had were kept for milk, cream, butter and cheese for the family and the hired help. Milk was kept in crocks to cool. After the cream rose to the top of the crocks it was skimmed off to be used as cream for cereal and cooking. The cows usually produced more milk than was needed by the family. The same was true for eggs from their chickens. Margaret's mother Mabel sold the excess cream, butter and eggs in town for groceries not raised on the farm. This money was kept by her mother for extra things. Money made from the sale of milk, butter and eggs was not a part of the farm income. In keeping with the practice of many farm family operations, this money was spent at the discretion of Margaret's mother.

Press A., Susanne K., Donald D., Rick K., Dotty R., Donald G., Wilma Q. and others told of similar sale of excess farm produce over and above family use. Even though farm women did not have much to say about the income from the sale of livestock and grain, they reaped the benefits of careful management of the products considered for home use.

The Huckster wagon is here!

Persons providing products needed and desired by farm families had more than one option in the 1930s and 1940s. Business could be set up in a store in town where they waited for farm families to come to town for their trading. Some enterprising persons took their products to the farm. Items for sale or trade were transported by a vehicle often called a Huckster Wagon. Initially the vehicle was not much more than a horse drawn canvas covered wagon. The inside was lined with shelves and boxes to accommodate a variety of items like cookware, small household tools and utensils as well as food condiments not usually home grown. In later years the huckster wagon was more of a bus. In fact some of them were renovated school buses.

Dot H. remembers groceries and other supplies being delivered by a horse-drawn huckster wagon. Many farmer's wives depended on it every week. They sold or traded eggs, chickens and butter in exchange for groceries and other items for the homes of a farm family.

The truth will be told

Impulse buying started long before the displays were purposely placed near the checkout counters at supermarkets. A savvy huckster man would try to arrive when children were home. Candy, a rare treat, was basic to the success of the huckster business. Gene T.'s dad Jack at one time delivered and sold products with a motorized huckster wagon. Money making was not assured. It didn't help matters to have five boys in the family. At times some of the boys got in the wagon and laid claim to their part of the family business. They took it out in the form of a little gum and candy. One of these times a son or sons, took the whole supply of chocolate Ex-Lax (Laxative).

When Jack discovered the absence of the candy counterfeit, he confronted the boys and said, "Alright now, who did it?" On not getting a confession from any of them, he concluded the confrontation with, "That's alright, I'll find out soon enough." It may have come as a surprise for them, to discover taste satisfaction was only one part of the benefits of Ex-Lax. Conclusive conviction was soon to come.

The bigger—the better

John Frazer, cousin of Bob Mc.'s, wife Kathryn, worked in his dad's grocery in Wilmington. One of his regular customers was a lady who came to get eggs, among other things. For most customers, an egg is an egg. Though some did prefer brown shells over white, the inside was much the same with egg white albumin and yellow yoke. As long as they weren't cracked, most customers were satisfied. One lady however, insisted on the largest eggs Mr. Frazer had. It was a bit inconvenient at times, but the customer is always right. If you are going to stay in business you put up with one and all. One day someone brought in some duck eggs in exchange for groceries. Frazer's Market did not usually buy or sell many duck eggs since they are nearly twice the size of regular hen eggs. John had an idea. You can guess what happened the next time Mrs. "largest eggs" lady came in. John gave her duck eggs and didn't tell her. After that he always gave her duck eggs if he had them. Bob Mc. thinks Cousin John never told her the difference; would you? On the other hand the lady might have known all along, thinking John didn't know the difference. Why tell and risk losing the "big egg" bargain? It doesn't really matter. It was a win/win situation. I liked the story. If you did we can chalk it up for two more wins.

We'll bring the chicken

Some hard working farmers found time to do a little loafing in the winter. A good place to loaf in Lumberton, Ohio was the country store. Sitting near the "pot-bellied" heating stove eating crackers and cheese, the men would share the latest news of the farm, in town and beyond. There was always time to talk of the weather and the telling of tall tales. Don G.'s grandfather Art dropped in from time to time. His grandfather made maple syrup in the late winter and early spring. It had been a good season with a number of days of ideal freezing and thawing weather. This morning he stopped in the store and told the guys gathered there he was planning to start the fire under the boiling pan to boil the sap to make syrup.

The guys sitting around the stove suggested it sounded like a good time and place to have a chicken roast. Don's grandfather agreed to the get-together on the condition that they'd bring the chickens. It was a done deal. That evening the gang showed up at the sugar making with eight chickens, killed and ready to roast. His grandfather helped them wrap them and put them in the coals. What a feast and a grand evening! The men ate their fill of fresh roasted chicken. They left after thanking Art for his generosity. Sometime later Don's grandfather found out the chickens were stolen; stolen from his chicken house. What are a few chickens when you have good friends?

How can we help you neighbor?

A hardware store in the small country towns was one of the important places of business for farmers of the family farms. Mike D. grew up working in his grandfather's hardware store in New Vienna, Ohio. Most of the customers had learned from experience that Mike knew the business. Not all of the ladies realized how much he was trusted by his grandfather to take care of the customers. When he was about nine years old Mike had something of a problem with one such farm wife. She had been sent by her husband to get a certain tool. Mike met her when she entered the store and asked if he could help her. Although it was clear she did not trust a young man to know much about it, she told Mike she was sent to get a file of some sort. Mike took her where the tools were displayed and showed her the choices. The farmer's wife grabbed him by the arm and in a huff led him back to the office to his grandfather.

Grandfather Estel asked her, "What's wrong?"

She was very agitated by now and said, "I will not stand having a young boy with a foul mouth around me!"

He turned to Mike and asked, "What did you say?"

"I just asked her what she wanted," replied Mike.

The lady interupted, "I told him I wanted a file."

"What did you tell her Mike?" questioned his grandfather.

"I just asked her if she wanted a Flat Bastard or a Double Cross file," replied Mike.

Mike's grandpa picked up the file box and showed the lady the file name was "Flat Bastard," but she did not believe him and left wanting to wash that young man's mouth out with soap!

I know just what you need

Dick G. does not like to go to drive-in-windows. He still likes the small town custom of going inside and talking to people. This is one of the reasons why Dick is known by most of the people where he does business. Going into the establishment affords more opportunity to visit and talk with the owners and employees. At times it may have to be brief in consideration for others waiting to be served. He recently went into the bank to deposit a check in his account. Dick admits he often goes in dressed in old work clothes. When he presented the check to be deposited to a lady teller who is a good friend, the verbal bantering would not have happened at a drive-in-window in most establishments. When he handed a check to the teller for deposit, she asked if he wanted some money back. Dick said, "No I just want to deposit it."

She questioned him again saying, "Are you sure you don't want some cash back?"

Dick replied telling her, "No, I just want to deposit it all."

The teller insisted, "Dick, you should reconsider. I do think you need to take some cash back."

Dick knew she was getting at something, but was not certain why she kept insisting that he needed some cash so he asked, "Why do you think I should take some cash?"

"I just think you should," she said with the hint of a smile emerging on her face.

Dick finally asked, "What for?"

The lady replied, "Dick you really need a new jacket."

Helping Hands and Feet
Help of hired Labor, Children, and Pets

"Helping out" defined the work of young children. Tasks were assigned to young farm children as young as five. Children worked in the home or barn, and women took responsibility for teaching these chores. Young children milked, brought in wood, helped in the garden and with chickens, herded cows and did errands, such as taking water to field workers.
Mary Neth—*Preserving the Family Farm*

While some thought the term farmer applied mainly to men as owners of the farm, in reality others were also farmers. Women, children and hired helpers were farmers. Not only did they have a vital role in the whole operation of the family farm, in many instances women and children took an active part in livestock care and the operation of farm machinery. Additional help was both seasonal and part-time. Farm pets such as dogs and ponies figured into helping with livestock herding and transportation for children with errands to run for the farming operation.

Most of the people interviewed have high praise for being raised on a diversified family farm and knew their children a generation later felt the same. Their lives were enriched by a wide range of disciplines from chores around the house to working with a variety of animals and poultry as well as in the fields. Farm youth not only developed many marketable skills, but had a broad understanding and appreciation for life in all its forms as well. Children learned in the laboratory of personal family farm experience the wonder of the birth of a newborn to the reality of the sacrifice of prized livestock for food and financial gain.

Children—their labor of love

How does your garden grow?

Garden work was a matter of choice for Jim, David and Becky H. However, the choice was their mother Mary's. A garden was a "must" for feeding the family. Working in the garden was a "must" for the children. As a teacher of children in a public school, Mary had her own laboratory for learning and instruction for her three in the family garden. Labor lessons learned and appreciation of the God-given gifts of the garden has served them well as adults. Ever the teacher, Mary was specific in her job assignments. Jim especially hated to be assigned rows that were his responsibility. He knew there were other things he would rather be doing like driving the tractor or riding a horse. As an adult he took up gardening and enjoyed it. Jim is thankful for the early experience gained in spite of his strong dislike for that manner of child labor.

All three of Mary and Bob's children as adults enjoy gardening. They all share their parent's work ethic. Thanks to mother Mary, home grown food continues to be important for many others who have received gifts from her garden.

A lesson learned

Teresa H. and a cousin Kenney felt good about helping their grandmother when pulling weeds in her garden. They loved Grandmother Julia and this was something they could do. The children were tired of being told they were not big enough to do this or that. It seemed like a lot of weeds when they started, but they kept pulling them up and made a pile of them to show their hard work.

It would be nice to give this story a happy ending. I think I can but first I must tell you what else Teresa told me. Mistaking them for weeds, these sweet, enthusiastic, young children pulled up about a hundred tomato plants their grandparents had planted. Teresa and Kenney lived to tell the story is the obvious good thing about it all. The love of grandparents remained strong and the grandchildren learned one of the many lessons of life that must be seen through tears.....

Money earned

Dorothy D. as a young person walked across the fields to a neighbor's strawberry patch and picked strawberries and was paid 2 cents a quart for her work. When she took the money home Dorothy put it in a jar. It stayed there until they went to the store in Lees Creek, Ohio. At the store she could spend her hard earned money on ice cream. Ice cream cones sold for 5 cents a piece. It is good to keep money in circulation.

Chickens—from peep to peck

At a stockyard the commotion of complaints of confined livestock is a given. If you are bothered by the loud moans of cattle, the grunts and squeals of swine and the bleating of lost lambs; don't go there. Today we do not expect to hear much noise in a post office. A few people remember the arrival of spring being announced in country post offices by the frightened peeps of hundreds of baby chicks. Every spring Dot H.'s mother Myrtle ordered 500 baby chicks. The chicks were delivered by mail. Chick hatcheries had special cardboard boxes with ventilation holes that held 100 cuddly noisy peepers. Hatcheries usually sent a dozen extra chicks since some might not survive the trauma of the trip.

After the chicks were picked up at the post office they were put into a small heated building. Once in a warm place the baby chicks were taken out of the five boxes one by one and given a drop of water with a small syringe. Dot was impressed by how quickly her mother held the peeping chick in one hand and slipped a welcome drop of water in the little open mouth. Dot's job was to go out in the brooder house and see that they had feed and water, but she was not old enough to put in the feed and water. Her chore was just to check and see when more feed and water was needed. The reward for her labors in addition to the pleasure of seeing the babies grow was a dozen baby chicks of her own to raise.

Laying hens sat—when pestered they pecked

Cuddly soft baby chicks at times grow up to be rough and rowdy roosters or mean old mother hens. Don G.'s mother Edith had about 400 hens. Don found it was no fun gathering eggs from under hens sitting on their eggs. Many of these hens did not know the rules of poultry management. The hens had the mistaken idea the eggs were theirs to keep warm and become baby chicks. The hen's, "Oh no you don't," took the form of a menacing look and a mean peck on the hand that dared to rob them. When he was younger Don didn't like to hear his mother tell him it was time to get the eggs.

One of Rosalee W.'s chores was to get the eggs from the chicken house. She had no problem when the hens were not sitting on the eggs in the individual straw lined cubicles. Like Don G., she had been pecked by hens who did not like to have someone messing around with their eggs. She'd had enough of this pecking from pestered hens when she remembered the roost scraper. She took it down from where it hung near the place of nightly rest and relief for the hens. Rosalee figured that would be just the thing to lift the nasty old biddies up off the eggs long enough to reach in and rake the eggs out. Then she could retrieve them without the surprise of a skin bruising beak. For some reason her mother Grace took a dim view of this method of egg gathering when the scraper broke more eggs than it saved.

Margaret H. as well, was relieved when she found hens off the nest because some hens did not want you to touch the eggs. They were quick to let you know.

Not much to crow about

Jim H. told us on their farm they had Red Rock chickens and two white Leghorn roosters. His grandmother had Bard Rocks and White Rocks. Of all these feathered friends, one clearly stands at the bottom of his list. It was the smaller of the two Leghorns that was feisty and would sneak up on you from the back. Jim had no love for that rooster.

Sue H. also remembers a rooster that would peck and scratch you on the legs when your back was turned.

Rooster meets his match

In the summer Susanne K. had the choice of doing the dishes or giving the chickens their noon feed and water. She always chose the chickens. Her grandfather Arch H. had a small barn on their place at Grassy Run. He kept one cow. Susanne always like to be with her grandfather and came out one evening to watch him milk the cow. As she stood watching, Susanne was suddenly surprised by the attack of a scrappy rooster that sneaked up behind her. In an attempt to let her know he was boss of the barn, the rooster flopped and scratched her legs. When Susanne screamed, her grandfather jumped up off his milking stool, left the half-filled pail of milk by the cow and came to her rescue. The rooster was so intent on chasing this young intruder out of the barn, it did not see Susanne's grandfather in time to escape the swift grasp of his strong hands. We don't know the rooster's take on the tale since Susanne's grandfather grabbed it by the neck and pulled off its head. Taking it to the house he called to his wife and said, "Maggie, you have a chicken to dress."

Not for money

Rick K.'s chores were to take care of the chickens. He had to feed and water them and gather the eggs everyday. Rick enjoyed the work with the chickens, but never got paid for his chores. It bothered him that they were his chickens to care for, but when the eggs were sold they were his parent's chickens and they received the money for the eggs. As an adult Rick has a better appreciation of the intangible return he gained from chicken chores.

Mary Lib S.'s girls went to the barn almost as soon as they could walk. They started feeding the calves and grew up in the barn loving and caring for animals.

No, you may not milk the cows

Dot H. admits she was a "Tom Boy" and is rather proud of it. She loved to go out on the farm and be with her dad. On the farm they had 12 milk cows. Dot's mother Myrtle milked the cows. Dot really wanted to help milk the cows, but her father would not allow her to do so. Dot was told she could put the feed in the trough for the cows, give them their hay and shell corn to feed the chickens. There was no changing it. Dot could not milk the cows.

Have you any wool? Three bags full

Jim and David's Dad Robert H. sheared sheep. Wool from shearing was put in large sacks. The boys had the job of getting on top of the wool after it was pushed into the sack. Jumping up and down on wool in the sack was fun. However, having fun at work was an extra. The purpose was to help compact the fluffy wool to get more in each sack. They enjoyed this recreation to their labor as well as being with their father and helping him with the work.

All day for half a buck

During the depression of the 1930s Press A., as a teenager, lived in town but worked for farmers whenever he had the opportunity. On his first paying job Press drove a team of horses as he sat upon a seat directly above a roller. The roller was used for crushing dirt clods and packing the ground. Loose powdered dirt helped keep the soil moisture from evaporating on the hot dry days of the 1930s. All day long in the heat and dust Press drove the horses in a big field planted in corn. His pay for this day of labor, besides a sun tan and challenged muscles was 50 cents. He gave it to his mother because they needed all the money they could get.

Help for hire—an extra hand finds a home

Diversified family farms often needed more year-round help. Those who worked fulltime were referred to as Hired Hands or Hired Men. In some cases women and girls were hired to work in another's home in times of special needs. Year-around hired help with a family, was usually furnished a house, some food produced on the farm and a small amount of money. Some men starting as hired hands were interested in moving up "the agricultural ladder" to eventually rent or own a farm. Others were content to have a place to live and raise a family without the headaches and obligations of ownership. It was a satisfying life for all when the hired man and farm family had a good working relationship.

A good job for the hired man

Don G.'s parents had about 40 beef cows. One of his chores was to go up in the silo and using a pitchfork throw down ensilage for the cows. Don did not enjoy this chore and was glad to let the hired man do the job when he offered. The hired man that worked fulltime for them was provided a house, a gallon of milk a day, two hogs a year at butchering time and paid $13 a week. They were happy to have a hired man who stayed for awhile. Some hired men and families often changed from farm-to-farm.

When "I will" exceeds "I will not"

The "hired hand" who worked for Ray C.'s father from 1937 to 1951 was a bit unusual. He did not drive a car, did not like to drive a tractor and did not milk cows. It would appear he was not a good choice to work on a dairy farm like Ray's. The man did however, have some pluses for farm work. He was good with horses, a hard worker, all around hand for field work, worked in the garden, mowed the lawn, cleaned the fence rows with a scythe, built fences, was a good painter and worked in the woods. Ray's family had 4-5 acres of cucumbers for pickles in the 1940s. The hired man did a lot of the work, planting, hoeing, cultivating and picking the cucumbers. He worked in exchange for house rent, milk, firewood, garden produce and two hogs a year, plus $16 - $20 a week. Ray's family had a good friend as well. Who cares if he didn't drive a car or tractor? You can always find some early rising, soft-hearted guy who enjoys milking cows, can't you?

Between the two, the work was done

Margaret H.'s father Alton had hired men to help so she did not do much of the farm work with livestock. One of the hired men Ed, came to the farm in the 1920s where he and his wife lived and raised five daughters. Margaret had no siblings, but was blessed by the friendship of these girls. Ed worked for a house, a garden, and received the customary meat, milk and eggs plus a modest salary. Later Margaret's father Alton needed more help and Ed's brother Orville joined them and received produce plus wages for his labor. Both men and their families were grateful for the arrangement and had no desire to climb the "agricultural ladder" by renting ground on their own. Starting as a "hired hand" was for some, the first step to actual ownership of the land. Ed and Orville lived out their lives satisfied to be secure as dependable "hired hands," good neighbors and community members. It was a good life, and they did not want to leave it. Both men stayed on the farm until death. While they were not blessed with monetary abundance, or ownership of land, they were family farmers blessed with the same good life of those who owned the land.

Other hired to help

A good break from prison camp

After dinner they would wash their plate and stack it on a pile of clean plates, then go by and thank Burdette Q.'s mother for the food. They were especially thankful for the pie she made for them. These men were German prisoners captured during WWII whose camp was in Wilmington, Ohio. Burdette and Bobby Q.'s father Frank had arranged for prisoners from town to help shock corn. The prisoners were guarded with a gun but gave no hint of trouble. The men were good workers. Racks were put on the farm truck to pick them up and take them back to town. It was a good working relationship. The corn was shocked and prisoners got a chance to get out and have a delicious meal. They did not have much to eat in the prison camp. This simple sharing of labor for good food, in a small way helped bridge the gap between two nations at war.

Oh no, don't tell me he's back

Wilma Q. remembered a man from out of state who came back to her parent's farm year after year at the time of shocking corn. He could keep up with the best of them in the cornfield so he earned his pay. When he worked in the corn on their farm, he had meals in their home. The man was nice enough, but he seems to have had an aversion to bathing. Their eyes did not always tell of the problem each time he came around, but their noses never failed. After his departure, and only then could Wilma's Mother Emma breathe a sigh of relief.

What's this thing you have with weeds?

Eleanor G. was again on a family farm after her marriage to Phil, a Clinton County farmer. During the seasons that required more labor they had boys who worked on the farm. Some boys lived with them. Phil had high expectations for these young helpers. No matter what the job, whether pulling weeds out of the crops or working in the heat of the hay harvest, Phil was their mentor and friend. Not all appreciated his high expectations of labor at the time, but later in life most realized the true value of working on Eleanor and Phil's farm. These young hired laborers became close friends. As adults they came back to the farm for a visit from time to time, ever thankful for their friends and all they learned while working on a family farm.

Helping pets and the part they played

A pony was affordable recreation for family farm children. Ponies often proved to be agents of learning for children as well. As with other pets, children learned both the pleasure and the responsibility of caring for animals. At other times children used a pony and their riding skills to run useful errands such as taking water to the thirsty men working in the fields on a hot day. When children cared enough to be responsible they learned to understand and cope with the character and nature of their pets.

Reigning cats and dogs

What can you do with a boy or two?

They had only a few cows on their farm. David H. and his Brother Kenneth each had a cow to milk. After a boy has mastered the art of milking a cow, there is little to challenge the boredom of the routine. It is good to be working with another who is just as bored. If the cow does not break the routine by upsetting the milk pail or kicking, one of the boys finds a way. David and his brother got into milk fights using the cow's teats as squirt guns. After calling a truce the boys fed the barn cats the same way. The cats got up early to watch the fun and lap up the warm milk.

The work of a good watchdog

A boy needs a dog growing up on a farm. The next day after he was born, Rick K.'s father Alfred bought a dog. He felt the time was right. A boy should have a dog so the two could grow up together. As his parents hoped, the dog grew to be very protective of Rick. That was good because, Rick as a child was inclined to wander out among the livestock. Knowing the dog was always by his side at such times gave Rick's parents more assurance of his safety. Rick had a pet lamb that followed him around like it was a dog. His dog seemed to think the lamb was just another dog. It worked for them. Rick had protection and the pleasure of two pets.

When Rick was about 8 years old his dog was going blind. While chasing a rabbit the dog ran full speed into one of the tractor wheels. Rick's faithful watchdog did not live long after the impact. When they got another dog it was a little rat terrier they named Bimbo. The little dog, true to his breed was a big dog when it came to catching rats. Dogs do not live long enough to outlive the rats. Rats abound with grain on the farm. The death of Bimbo was followed by another rat terrier given the same name. Over a few years they had Bimbo 1, 2, 3 and 4. Each dog proved to a worthy watchdog, a personal pet and rat eradicator.

The watchdog is also watching you

Dot H. had no brothers or sisters until she was 6 years old. However, she was not without a playmate because her parents bought her a puppy. The puppy soon became her constant companion. As a budding teacher, her dog was the student. Dot taught him to join her at tea parties where he sat at the table and drank water out of a cup.

When the puppy matured he was her designated watchdog. Dot loved to be outdoors and was prone to wander beyond safe boundaries. Her dog did not disappoint her parents, but she was mad at him when he barked for adult help. If she climbed higher in a tree, or went out of bounds too near the creek her playful puppy, now grown guard dog would bark at Dot until she retreated or a parent came as his backup.

When dogs get in trouble

Dorothy M.'s family had two beautiful Collie dogs on their farm. The two friendly pets were helpful rounding up the livestock. At times dogs give way to natural instincts to hunt and kill. It is a sad day when the domesticated training to care for livestock turns to killing. Sometimes dogs with sheep can be bad news if they have not been trained for sheep herding. When the news came telling of a neighbor's sheep being killed and the evidence pointed to their two Collies it was heartbreaking. The otherwise gentle Collie dogs had to be put down. Even though it is never easy to loose faithful dogs, that's what good neighbors do when dogs do damage to other's livestock.

The price to pay for pets

David H. remembers their farm pets were loved and given special treatment, but they were not human and did not involve costly care. They ate what was available on the farm. Few people purchased pet food or had costly veterinary expenses. Productive animals worth the money got the veterinarian care as needed, but their ultimate worth was always a factor that had a bearing on how much was spent on their care. Dogs that got into a neighbor's sheep had to be put down, even though they were someone's pet.

Taste tested insulators bite back

The restless pony was intent on pawing down the

old fence. Gene T. decided to put up an electric fence. He bought two boxes of insulators and laid them on a discarded old trailer in the lot. When he had to go get some nails to fasten the insulators to the posts, Gene told his Dad Jack they'd better put the boxes up to keep the pony from getting into them. His dad didn't think the pony would bother them, so they went to a hardware store for the nails. Gene didn't know the pony had taken note of the two boxes while they were away. To satisfy her curiosity the pony managed to tear open both boxes. What she saw in the first box looked like something good to eat. When she tried to bite one of the insulators it was hard and did not suit her taste. To her disgust the second box was more of the same. In the process of her taste testing, she managed to scatter the insulators all over the ground. That's when Gene and Jack came back into the picture. They figured the pony would have a different opinion of these insulators when connected to all the posts and wired to the electric. The pony did not disappoint them. She followed Gene around biting other insulators as he'd get them secured to the post. As soon as they finished the job and put the electric charge to the top fence wire, the pony decided to give one of the insulators a second taste test. When she put her head against the electric fence wire her winter hair kept the pony from getting a shock. It was a different matter when she tried to take a bite out of the insulator. An electric fence alternates between an electric charge, off briefly and then another charge. With her first bite there was no reaction, just the same old tasteless white thing. When the seemingly lifeless thing bit back she got quite a kick out of it. The pony raced around the lot to get away from those white things that looked good to eat, but oh what a bite! She blamed Gene for the whole thing and would not come up to eat while he was around for a few days. After a time the pony gave the fence the required respect and Gene was no longer the bad guy.

The More We Get Together
Neighborhood Joint Labor Projects, Threshing, Filling Silos, Schools, Churches and Farm Organizations

The arrival of the threshing machine brought a thrill to farm people not only because of its impressive size, but also because it signaled the arrival of neighbors, work crews, hard work and socializing.
Mary Neth—*Preserving the Family Farm*

Even though farm families were self-sufficient most of the time, some activity called for shared labor. Harvest time for grain and corn for silage was labor intensive and necessitated seasonal harvesting equipment. At such times it was in the best interest of all for farmers to help each other. Threshing and filling silos were occasions when a half dozen or more farmers with extra help went from farm-to-farm working until all had their grain harvested or silos filled. Since there was a noon meal to be provided at each farm, women and children from all the cooperating farm families were also involved. The net result was neighboring farm families enjoying work, food, and visitation for many days. At other times for road repair and snow removal, farm families came together and enjoyed their labors for the common good.

Like most people interviewed for this book, I was not old enough to be involved in the hard labor of threshing. Yet, all old enough to remember these harvesting times of cooperation, have vivid memories of great neighboring. Most of the adults did not hide the fact that they were enjoying working together and having as much fun as the children. We had only one team of horses on our farm when I was able to handle a degree of adult responsibility. Harvesting grain for me involved a Massey Harris Clipper combine. This harvest machine was powered and pulled by a John Deere model A tractor with a power take-off. Without a cab it was dusty and noisy, but I loved it. The feel of power and purpose inflated my ego. I had every reason to believe this was about as good as it gets. At no time do I remember thinking, something is missing here. In the 70 years since that time, we may have a better understanding of what has been lost to larger fields of grain and vanishing tree lined fence rows.

In the early 1920s when the combination of the grain binder and thresher/separator came off the drawing board, few farmers had a clue to its significance for the American family farm. In about twenty years it became the instrument of choice for harvesting grain. It was a costly investment for a machine that would be used perhaps one month out of every twelve. One farmer with a combine could do the work of fifteen to twenty men at the time of harvest. Little consideration was given to the toll taken on rural cooperation and community. It not only replaced a number of men it meant a like number of women and some children. The combine may have put an end to some hard labor, but proved to be the eventual demise of a great time of rural bonding and the satisfaction of working together for the common good of farm families.

Farm families come together to get it done

With heartfelt thanks for fair weather and the promise of a bountiful harvest, the congregation of Pilot Grove Friends Church sang with renewed enthusiasm the Harvest Hymn. As a "pre-schooler," my immature ears heard "We shall come rejoicing, bringing in the sheep." At that point in my progress of becoming a junior family farmer, my brain had not found a home for the word "sheaves." While others were gathering "sheaves," I continued bringing in the "sheep." After being introduced to the binder that cut the wheat and bound the sheaves I came to unity with the others singing the words of the Harvest Hymn.

Harvest Hymn chorus– "We shall come rejoicing bringing in the sheaves."

Robert Mc. views "threshing" as one of those activities that made a community a community. Neighbors all worked together helping each farm family with their harvest, usually wheat and oats. He has vivid memories of the peak of excitement when he first saw that huge steam engine of threshing power belching thick black smoke as it came slowly up the road pulling the thresher/separator.

Mary Emma H. remembered the sound of the steam engine whistle as it came up the road. Neighbors came with flat-bed wagons for loading sheaves of wheat or oats. Others had box-bed wagons to haul the grain.

Even though farmers gloried in their independence, they found a degree of harmony as they worked together in harvesting grain. After a few years of experience men and women established a routine ritual of the procedures from gathering the sheaves to how the table was to be set for the noon meal.

Women and children were a vital part of the threshing experience, but were not involved in the operation of the machinery and intensive labor. Some women like Norma L. were active in the preparation prior to threshing. As a teenager, she learned to drive a team of horses. She rode the horse-drawn binder that cut and bound the wheat and oats. Stalks with grain filled heads were tied into bundles to be put in shocks.

David H. and others agree on the procedure for picking up the grain in the field. A few weeks before threshing, bundles from the binder were put in groups called shocks. Each shock had nine bundles. Four were lined up in a row. Two more were added to each side with one put on the top as a cap. With this arrangement the grain got more air circulation for drying in preparation for threshing. When the shocked grain was ready for threshing, one man or two with a pitchfork put them on a flat-bed wagon drawn by a team of horses. Another man, also with pitchfork in hand, known as the "loader," stayed on the wagon. His job was to stack the bundles on the wagon with the grain heads out and cut ends in. The loader also stayed with the wagon and fed bundles into the thresher.

Donald D. had stories to tell about threshing and the big machine at rest in the barn lot most of the year. The large silent separator does not give one a clue to its potential for noise. The cry of stones being crushed on the gravel road leading to the farm, are but a hint of the noise to come. The usual quiet of the country backs off when the separator is belted to the power of the steam engine. Wagons loaded with bundles were often pulled up to each side of the separator. It took a little time for horses to get used to pulling up and standing still next to the roaring rumble.

Flat-bed wagons loaded with bundles are on each side of the threshing separator. While horses patiently wait, a man on each wagon pitches bundle after bundle into the machine's cavernous mouth. As soon as one wagon is emptied, another takes its place. Wagon after wagon is pulled up to the mighty machine. At the other end of the thresher/separator the threshed grain was augured into a waiting box-bed wagon. Straw separated from its grain is blown out at some distance from the separator forming piles on the ground that will become a large stack by the end of a day of hard labor.

Cooperative activity a bit different in nature, but just as important to the success of the harvest was going on in and around the family farm house. Preparations were being made for the noon visitation of a host of hot, hungry and thirsty men. They were the honored guests to be served the fruits of the forenoon labors of women and children. An old man or two lingered around safely out of reach of harm's way. They found it hard to accept the fact that their manual ability no longer matched their memories of labor when they were active participants in the harvest.

Mary Emma H. remembers there were about as many women working preparing the threshing dinners as men who came in to eat at the welcome noontime blast of the steam whistle. Even though those interviewed were not involved in the same threshing rings, the stories they tell are very much the same. As children, they too were drawn into the excitement of the day's activities. The children cherished the confidence afforded them as they were put to work with the food such as peeling vegetables and fruit, setting the tables, putting up the chairs and preparing stations for washing.

Susanne K. and her sister were in charge of getting the table and wash tub set up for the men to clean up before dinner. They got the tub out and filled it with water early so it could be warmed by the heat of the morning sun. On the table they added a mirror, comb, soap and towels.

Dotty R. remembers putting the wash pans out on a stand near the pump. She was not surprised to see some of the men pumping cool water over their hot heads. However, Dotty did not expect to see them all drink the cool water out of the same cup. Had she been near the barn or out in the field, she might have had a problem with men passing the gallon jug from man to man.

Anne E. told of making tea for threshing dinners in a 10 gallon milk can. Dotty M. shared in the excitement of the occasion with a sharp eye to all the activities. The crackle of chicken frying and the aroma of fresh baked pies foretold of a great meal. Women had their specialties such as bread, cake or pie. In some threshing rings there was competition among some of the women to be the best cook. The men in charge of the rotation took note of the best cooks and tried to schedule the threshing so that they would be working on that farm at the time of the noon meal. After washing men ate first. When there were more men than could be seated at one time, the women prepared for a first and second shift.

When the men were seated at the table, Dorothy D. had the job of pouring coffee. Coffee was to be poured in cups with saucers. She had the misfortune of spilling some of the coffee into the saucer for a man who was not very forgiving or understanding. When he spoke to her with harsh words of rebuke, she went back into the kitchen in tears. Dorothy was comforted by her grandmother who said, "Don't let that bother you, they are tired. Just forget it." And she did, until now. What a wonder when we are blessed to remember what we forgot years ago!

As a young person Sue H. helped her mother with the threshing dinners. When thinking about that experience Sue remembers a man who always took his teeth out and put them in his glass before he ate. Women and children were the last to eat lunch after the rush was over. After a time of visiting and relaxation, the women and some of the older children cleaned up and made plans for the dinner tomorrow on another family farm.

A horse will do, if you don't have a pony

In the middle of a hot summer afternoon, nothing brought a smile to the face of thresher men like the appearance of a young girl and pony cart. A boy with a horse was OK, but the men had more good-natured fun teasing a "water-boy" that was a girl. Taking water to the men working around the separator and in the field was a high point in the celebration of the threshing harvest for children old enough to be responsible for the job. Mary Lib S. was proud to have been given this opportunity when visiting relatives in Iowa one summer. She was given a horse to ride to fulfill the duties of a water carrier. At the age of ten, Robert Mc. was a trusted "water boy." His duties included caring for his father Arthur's horse that pulled an old buggy. Robert delivered water to the men in one gallon jugs. Burdette Q. remembers carrying water for the threshers in one gallon crockery jugs wrapped in burlap with corncob stoppers. Cold water was allowed to spill and soak the burlap as it was pumped from the well to fill the jugs. Water evaporating from the heat of a warm day kept the water cooler. "Real men" could slip their index finger in the hole of the handle near the top of the jug and then with a one handed tip a heavy full jug to their mouth would drink before passing it on the next thirsty man.

Dorothy M. and her brother had a pony they used to take water to the men at the threshing machine. They rode on a cart pulled by their pony carrying a ten gallon milk can of water and a tin cup. Before drinking some men stopped to wipe the sweat off their forehead with a red or blue bandana. Each man took a drink of water from the tin cup and passed it on to another until all had a refreshing drink.

Susanne K. a true farm girl, who loved to work outside the house, had a pony and cart that she used to carry water to the men threshing. She followed the ring on its circuit from farm-to-farm a few summers as the water girl. She made enough money the first summer to buy a croquet set, a very popular game for young and old alike.

Much to her sorrow the water service was subject to a change. For some reason the pony was in a hurry to get back to the barn. Though she tried her best, Susanne could not rein in the pony. Susanne, pony and cart went around the corner at the gate cutting in too close to the post. One of the cart wheels broke relations with Susanne, pony and cart and elected to stay with the gate post. Her cart was out of service with no harm to Susanne or pony.

Susanne had to continue her water service without a cart. Water was carried on the horn of the saddle in a jug wrapped in water soaked burlap. As she rode the pony back to the working men, it was a bit uncomfortable with the wet burlap rubbing her legs, but she kept her job as "water girl."

Time out for pranks

At times such as threshing, there were a few jokers with good natured pranks. Donald D. recalled the time when one man took off his gum-boots before going into the house for lunch. Another who chose to be anonymous filled them with water and put the boots up on the platform at the top of the windmill.

Mike D. remembers the story of a man working with others at a farm where it took a few days to finish the harvesting. This man was not the brightest, but a good worker. Like a few other workers, he chewed tobacco. When they stopped for lunch most of the men who chewed tobacco spit it out on the ground. When leaving lunch they would reach in their pockets and get a fresh chew. This man without much money for tobacco said, "I'm puttin' my 'chew' on this fence post." When he came back from lunch he put his chew back in his mouth to get the most out of his tobacco.

After he did this for a few times, the others thought they would have a little fun with him. Instead of spitting their tobacco chew on the ground, a few men put their chewed tobacco along side his. The subject of the joke was last to come out after eating. When he stopped at the post, with a puzzled look on his face said, "I know it all came out, so I know it's gotta go back in."

Fire Stories

Monster machine torches a straw stack

The truth to the old saying, "Trouble comes when it is least expected," might give cause for worry all the time. Little good can come from living in constant fear of what might happen. Most farm families knew enough about the potential for accidents they greeted each day with a greater degree of optimistic caution than wearing worry.

Burdette Q. remembers the day when a wagon loaded with bundles of wheat was pulled up to the thresher. The men had just begun pitching the bundles into the mouth of the threshing machine when it happened. Suddenly on the top of the straw stack there was a big ball of fire. It seems a man carrying matches accidentally dropped some among the bundles going into the thresher. "Going into the thresher" made all the difference. The matches were ignited when they went through the machine. Straw was already on fire when it was blown onto the straw stack. Men of a neighboring ring of threshers nearby saw the fire and rushed over to help put it out. Total loss was 26 shocks of wheat and the straw stack. Quick action of the farmers by restricting the fire prevented a much greater loss. Consider the 260 gallons of gasoline in a tank near the intense heat of the rapid burning straw stack, and you realize the potential for a much greater threat to men and machinery. While some men with tractors plowed around the fire to prevent it from spreading, others took the belt off the engine running the threshing machine and pulled it away from the heat and flames. The team work of Burdette's father Frank and all three of his sons plus the help of many neighbors averted a major fire.

Left to do its own thing

Dotty R. lived down a long lane near Jasper, Ohio. Her father Millard, a school teacher, was away attending a meeting at Washington Court House, Ohio, when a fire was discovered in their house. Men working on a bridge nearby helped put it out. Dotty's father came home driving his Model T Ford. Coming down the lane he saw all the smoke and commotion. In a panic he jumped out of the Model T while it was still running. The car ran up against the fence and stopped. Millard didn't look back at the Model T as he ran to help put out the fire. Since the fire was nearly out, those gathered at the scene watched the humorous picture of the runaway car and Dotty's father's race to what he thought might be a major tragedy. It took him awhile to see how funny it looked.

The Model T and the fence were OK, since they saw no reason to hurt each other.

The table is returned

David H. and his family were eating supper when a fire was discovered. A whiff of smoke was their first clue that led them to locate the fire burning in the upper part of their home. Later speculation pointed to the possibility of a spark from the heating stove igniting the roof shingles. When they saw the smoke and flames, neighbors rushed over and helped get most of the furniture out. David remembers seeing his father Maynard bring out a table only to turn around and throw it back into the fire saying, "I never did like that damned table."

Help in time of trouble

A tree Bob H. was cutting down split and knocked him down and broke his leg. He ended up in the hospital in traction until he was able to continue letting the leg heal at home. Jim, David and Becky H. all were grateful for the consideration of their neighbors. It happened during the harvest season. Bob's neighbors brought their equipment and harvested and stored his crops. This spirit of neighborliness fulfilled a need to work together. Helping a neighbor in need was the major reward for their efforts. Most members of family farms enjoyed the fellowship of getting together and were committed to helping neighbors with unexpected accidents, sickness or other catastrophes.

Other cooperative labors out foxing the fox

At a time when most farm families raised chickens, a fox in the chicken house meant a loss of chickens. Farm land with tree lined fence rows and other wooded areas provided ideal places of shelter for fox dens. Press A. and others interviewed remember organized neighborhood "Fox Drives." An area, perhaps a mile square, would be encompassed by a human fence. Men and boys armed with clubs were lined up side-by-side with a few feet distance between them all around the area. Men did not carry guns for fear of the loss of human life. One of the men would set off a stick of dynamite as a signal for the drive to begin. They all began closing in on the foxes. Men and boys walked together shouting and making noise to rouse the fox. As the line closed ranks, there was little room for any sly fox to escape. Death by clubbing was not a pretty sight, but neither were dead chickens on the farm.

Filling silos—winter feed for cattle

Cattle farmers, beef and dairy, formed rings of cooperation much the same as grain threshing rings. Ray C. and his wife June were dairy farmers who joined a ring of six to eight other farmers and their wives. Men and women had a great social time working together. Since silage corn was harvested in the fall, not many children were able to participate in the activities. Usually it took only a day or two to fill a silo. Corn for silage was cut when the ear was filled and before the stalk was dry. Men often started early in the morning since they did not have to wait for the corn to dry. This meant a long hard day's work. Corn was cut in the field and hauled to the chopper by horse drawn flat-bed wagons. The chopper was basically an 8–10 foot long open horizontal conveyer belt that carried the stalked corn into an enclosed circular cutting blade about 4 feet in diameter. The blade cut and blew the chopped corn pieces up a vertical pipe to the top of the silo a distance of around 40 feet. Power to run the conveyer and turn the cutting blade at top speed came from being belted to a tractor.

Prior to the day of filling a silo the chopper and pipe had to be in place. The distance between the chopper blower vent and the top of the silo varies from silo to silo. Most were in the 40—50 foot range. Sections of twelve inch diameter steel pipe were put together on the ground. Enough sections were bolted together to reach over the top of the silo. The top piece was a curved neck that shot the chopped corn down into the silo. Once the pipe was assembled including the neck piece, it was pulled upright by a rope threaded though a pulley at the top of the silo. That's where the "fun" came in.

Someone had to climb up the ladder on the other side of the silo and get across to the point where the neck of the pipe is to be attached so the chopped corn would be blown inside to fill the empty tubular void of the silo. The foolhardy or courageous man is up there above a 40 foot drop to thread the long rope through the looping pulley. For Margaret H. this was one of the risks she dreaded for her husband James. When he was safely down on firm ground she'd breathe a thankful sigh of relief. If it was a covered silo, there was a narrow wooden platform just under the roof that must be crossed from one side of the circular silo to the other. It is best not to look down while crossing if you have any fear of height. On a silo without a roof about the only option was to scoot around straddling the 8 inch wide circular wall. The thrill is doubled with the same 40 foot drop both inside and out. When the courageous one reaches the point of the pulley for the rope, the rope is fed through the pulley and dropped to each side. One end of the rope is attached to the neck of the pipe. The other will be used by men below who pull slowly to stand the pipe up so that it can be safely secured over the top of the silo. The pipe will then be bolted to the spout of the corn chopper by the men below and be tied down above by the man on top. Now all he has to do is to turn around someway or scoot backwards to the ladder for the descent to the ground below. All is ready now for the corn bundles as the chopper comes to life with a loud roaring hum of the huge circular chopping blade.

Happy is the school age child when silo filling is at his/her home on Saturday. If he or she is awake early that day, they will be greeted with the announcement of the great day to come by the metallic cry of steel wagon wheels crunching gravel on the road to their family farm. What a great day. Most of the rest of the day they will be treated to the loud hum of the chopper wheel. All is well when the loud hum of the chopper is subdued by the clatter of chopped corn as it hits the fan of the wheel. Add to this the bumps and thumps of the golf ball sized pieces of corn banging along the sides of the pipe on their way to a free fall into the empty, but soon to be filled silo.

At noon the chopper noise is silenced. The horses are watered and tied. Men come into the farm house and join the women who have prepared a delicious hearty meal. Food, fellowship and a bit of rest and the men are back on the job hoping to get finished in time to go home and tend their evening chores.

The women clear the table, wash the dishes and make plans for silo filling at the next place on the ring. As to be expected they enjoy a bit of friendly neighborhood gossip. It is seldom malicious. They simply care for each as part of a great family. The men will be doing the same hard work all day long at the next farm. Boring routine? Perhaps, but the farmers are not inclined to see it as such.

From river-bed to road-bed—neighbors come together for road repair

Rural roads were put to the test as farm families became more dependent on businesses in the neighboring town. The businesses in turn were more dependent upon their sales to the family farmers. Gravel roads were a marked improvement over packed dirt that was satisfactory to a degree when it was not dry and dusty or wet and muddy. Rural roads and their upkeep took on greater importance to both the farm families and their neighbors in town, especially after Rural Free Delivery of the mail. Even the much improved gravel roads needed at least annual repair due to greater use and the forces of nature. In the few times when the demands were not as great for their time, farmers worked on the roads. Back then equipment for road repair was not complicated, but farmers were better equipped to do the job. Instead of leaving it up to each farmer to do what might be considered his share of the road, farm men and boys worked together. Robert Mc., Press A., Bob Mc., Mike D., Donald G., and others all told much the same story of the process of road repair. It might be considered a 3 bed process, from river-bed to wagon-bed to road-bed.

Basic to the process were horses, wagons and manual labor. It was hard work for both horse and man, but with minimal financial outlay. There was no charge for the river washed gravel. Horses were fed from the produce of the farm, and the minimal pay if any, was mostly incentive for the young boys. While motivated to work for the benefits they'd receive from improved roads, the men and boys were rewarded by the satisfaction of working together. If pressed for the truth, men and boys would not deny the motivation of competition. They took pride in the skills of their labor and the merits of their horses. Experienced men knew enough not to load too much on the wagon for fear of getting stuck in the mud of the road and sand of the stream.

The wagons exhibited men's ingenious understanding of the task at hand. Wagons with low side boards for ease of loading were horse-drawn down to the river or creek to be loaded with gravel. The wagon beds had 2 inch bed boards that could be pulled out from the side one board at a time. By this means as the horses pulled ahead, the men were able to spread the gravel out on the road rather than dumping a whole load in one heap. This saved hours of having to shovel and spread gravel around by hand. At times they took the wagon to the creek to soak the wood of the wheels so it would swell up to be tight and keep the metal wheel rim from coming off.

In the winter when they had a heavy snow fall, Robert Mc. joined his neighbors and some of the school boys to shovel the heavy snow the length of

Clark road by hand. In the summer they hauled gravel and spread it by pulling out the bed boards one at a time. Gravel was taken from the nearest creek. It had to be done every 2 to 3 years. Putting new gravel on the road was always a community project of volunteer labor.

Burdette Q. worked clearing the snowbound road in his neighborhood. The one he remembers most was a major snow storm with deep drifts. Burdette and his neighbors started shoveling the snow on their end of the road while neighbors on the other end did the same. The adult neighbors worked without pay because they needed to have the road open. Young boys like Burdette got 50 cents an hour for their part in shoveling the snow so the road would be passable again. All who worked reaped the satisfaction of a job well done in the company of friends and neighbors.

Under cover work

Margaret H. remembers women coming together for quilting as a time with social and practical benefits. Young women learned sewing skills and artistry from older women in the group in a social setting.

When women gathered for quilting in Kentucky, young mothers brought their preschool children with them. The quilt frame as Gene T. recalled, was suspended from four hooks in the ceiling. At times even children were involved in quilting. It was something of a game for them. Children playing underneath the quilt would push the needle and yarn from underneath back up to the women working on the top. It was a time of recreation and socialization for the children as well as the women.

Schools—a center of community

Rural people were especially wary of school consolidation. The one-room school was an educational agency, of course, but to rural people it was also a social institution that helped unite neighborhoods, uphold their values, and express their identities.

David B. Danbom—*Born in the Country*

Donald D. began his education at the age of 5 in a one room rural school. He chuckles when claiming he was always in first or second place in his class, a class of two. The school had one teacher, eight grades, a genuine slate black-board with white chalk, but no indoor plumbing. When they were thirsty the children, with the teacher's permission, could go outside the building to the pump. Everyone drank out of a tin cup that hung on the pump. Donald went from this humble beginning to become a successful farmer, school teacher and administrator with a Master's Degree in Education. He has lived through the years of significant change brought about by school consolidation. School consolidation has resulted in a better education enabling rural children to be better equipped for the demands of a changing world. However, most of those interviewed remember the unity of neighborhoods that centered in activities at the school. Teachers came from or quickly gained an understanding of a rural background. School board members were from the neighborhood. Mothers helped in the special activities such as school plays and celebrations attended by most of the people in the area serviced by the local school.

Donald G. attended Port William School at a time when children did not have to bring their lunch from home. Port William School had a Home School League responsible for providing meals for the children. In 1942 the League felt meal prices had to be raised from 12 cents to 20 cents. Donald's Mother Edith told them she could feed them for 15 cents. She got the job. Every day for a year they went into Wilmington to Sabins Market in their 1939 Plymouth to get the food for the children and brought it to the school in Port William.

Tom W. remembers the last day of school each year ended with a picnic. The mothers of the school treated the children and teachers by planning and providing the picnic. It was a great way to start summer vacation.

Do You See What I See?

Rosalee & Tom W. and Mary Eleanor H. went to school in Port William. All 12 grades where in one building. Grades 1- 8 were in the lower level. High school was on the second floor above.

Sue H. rode the school bus and enjoyed the time being with other students. She took her lunch in a black metal lunch box. Sue and some of her friends liked to sit on the back seats of the bus. They enjoyed the mini rollercoaster effect as the bus went over bumps. She remembers one time when the bus hit a large bump in the road and she bounced up and down. While the upshot brought a breath taking thrill, the downfall was something else. This time she came down on her lunch box bending it out of shape.

She was in for greater excitement once, when the bus driver took too wide a turn off Starbuck road onto Haws Chapel road. The bus ended up on its side in the ditch. Children in what became the upper seats slid down to join those below them. It turned out to be a jumble of arms and legs. Miraculously no one was hurt. This experience may have been the challenge Sue needed to become a bus driver. She was a school bus driver for the Wilmington City schools for 35 years.

At a time when many rural children rode a bus to school, children in town had to walk. Joanne G. lived in town almost a mile from the school. They did not have bus service. At noon Joanne walked back home for lunch. She had just enough time to wash up, eat and walk back to school for the rest of the day.

Those interviewed have fond memories of their school days. When they returned home each school day there was always a parent or another adult to meet them. They appreciated the bond between the school and the neighborhood as a relationship that contributed to their quality of life. Teachers were local friends and respected partners in parenting. When teachers found it necessary to discipline a child, they usually had parental support. Many schools employed a bus driver as the school janitor. Teacher, bus driver and janitor, each in their own way contributed to the education and maturation of children on the family farms and in neighboring towns.

Farmer's organizations

The new system of Farm Bureaus, Homemakers' Clubs and 4-H Clubs, created by the extension service in 1914, individualized the farm family by isolating its members. They defined the interests of men as agricultural and those of women as homemaking, mirroring the specialties of most experts. These organizations concentrated on the work of individuals in the nuclear family rather than the family and its role in the farm community.

Mary Neth—*Preserving the Family Farm*

The Grange founded in 1867 is the oldest of the "big three" farmers organizations. The National Farmers Union was founded in 1902 and the American Farm Bureau in 1919. They all were formed to improve the well-being of farmers and their families. For many reasons all three faded from the picture of farm families coming together on the local level. In our area the Grange was the last one to discontinue having a booth at the annual county fair.

Rick K. and his family have been active participants in the Grange. He knew it as an organization for the whole family. In the Grange, children were recognized as a vital part of the family farms. It also involved religious ritual in its meetings and membership ceremonies. The Grange is not as secretive as in the past. Young people attended Grange camps for rural life education, crafts and recreation. Children participated in the opening ritual of Grange meetings held in Grange Halls and other public buildings, township halls and churches. Susanne K. remembers selling food and crafts at ice cream socials and farm sales. Young adults had plays, talent shows and often performed at state and national conventions. Granges had display booths at the county and state fairs. The fellowship was prized even more than the money made. Grange was instrumental in getting FFA started. It was also instrumental in getting governmental help for other farm improvements such as rural electrification and farmers insurance.

Today the Grange does not have as much emphasis on the family farm and farmers as it used to. The Grange was a strong organization until marked decline in the 1950s and 60s. Rick K. thinks this change may be in part because of many more activities for children in the schools. Much like the Grange, other farmer's organizations, country schools and rural churches are closed or challenged by dwindling attendance and support.

Rosalee W. was in 4-H for many years. Her mother Grace was a leader. 4-H was different than it is now. Very few girls had projects with livestock. Girls had more domestic projects in keeping with the under girding conviction that girls and boys had different areas of labor on the family farm. There were exceptions in cases where there were daughters but no sons to help. At times girls chose to be involved with farm work in the fields and around the barn. In the last few years we find more girls involved with livestock and boys in cooking than 60 or 70 years ago. However, with the diminishing livestock numbers on family farms and a declining interest in home-cooked meals, 4-H will become more of an urban organization or cease to be.

Susanne K. acknowledges there no longer seems to be a need for FHA (Future Homemakers of America). With the availability of microwave "heat and serve" meals, ready-made clothing, laborsaving appliances etc. there is little interest in FHA today. FFA (Future Farmers of America) continues to be strong in some areas but with a limited number of students. Not as much emphasis of work with livestock, more on the cultivation of qualities of leadership. The few young people in FFA growing up on diversified farms are role models and our hope for the future of the family farm.

Why Pay to Play?
Games and Rural Recreation

Many older farm people considered spending money on recreation wasteful. A 1936 Illinois survey found that parents approved of church activities, visiting, parties, and fairs as recreation for young people. They found town-centered activities, such as street carnivals, dances, and "street loafing," objectionable. Youths, on the other hand, accepted the idea of leisure time and wanted these new forms of recreation.
Mary Neth—*Preserving the Family Farm*

A grandmother was telling her little granddaughter what her own childhood was like: "We used to skate outside on a pond. I had a swing made from a tire that hung from a tree in our front yard. We rode our pony. We picked wild raspberries in the woods." The little girl was wide-eyed, taking all this in. At last she said, "I sure wish I'd gotten to know you sooner!"

At a time when recreation is big business and financial gain, it may stretch our imaginations to think of having fun with little or no financial outlay. That was how it used to be. Those were times when youth and adults alike spent little money and personally participated in recreational opportunities. Even though you may have been one of the last to be chosen in group games, as was my case when they were athletic, you were expected to participate. Though often labor intensive, finding pleasure in their work was conducive to good health and happiness. Farm people who worked together found their labor at least in part recreational. With little or no financial investment, they netted their recreation as financial gain.

Creating our own imaginary farmstead is grounded in happy memories of carefree days. A "pretend" farmstead is best constructed in a grove of trees away from practical parents who have long wandered from the land of "make-believe." Here we find an abundance of construction material, sticks, stones, tree leaves, grass hay, etc. at no cost.

A call to adventure with mother nature

Creek or crick, take your pick

According to my 1966 Funk & Wagnalls dictionary, a creek is a stream intermediate between a brook and a river. A crick is a dialectical (peculiar to a certain particular region), of creek. I admit I am peculiar and come from a particular region in East-central Illinois. We are particular about a stream intermediate between a brook and a river. It is a crick! However, in deference to those who remind me that creek comes before crick in the dictionary, I will use creek, but I'm going to pronounce it crick.

In another book I wrote about "Peppermint Bay." I cannot actually take you to the Peppermint Bay that I knew as a boy in Illinois. We can look at the creek where it was, but Peppermint Bay is no longer there. It's only a memory now of a time when the small creek we can easily step across was a make-believe mighty river. A mighty river that carried barges of grain, cattle, coal and new farm machinery all made of sticks and weeds and corncobs. It was in a bend of the creek just south of the bridge. The cut bank here was high enough to slide down into the creek and be safely hidden from the unimaginative, workaholic adults. Across the creek cut-bank was a shore of very fine sand, ideal for making city streets, hills and adjoining farms. Peppermint Bay was peaceful and prosperous. There is no video game or factory-made toy that can touch it in terms of hours of pleasure and challenge to the imagination.

Susanne K., Burdette Q., Dick G., Rick K. and Jim, Becky and David H., are but a few of those who remembered the hours of great fun playing at a creek nearby. Jim H. played in a creek in front of their house. It was here beside his mighty river he launched little 2 inch toy figures on small board scraps that became cargo ships. Under his watchful eyes and guiding fingers the mighty ships found the lanes of safety between Gibraltar rocks. Ships of Jim's fleet floated on to pass "make-believe" cities on the creek's sandy beaches. At other points the ships found places to dock where their passengers could go ashore in search of adventure. By some feat of magic the mighty ships with all its persons and cargo were returned to a place of starting over again.

It was a special treat for Jim's sister Becky to go with her big brother to play along the creek. It was extra special the day that he allowed her to join in one of his favorite pastimes of watching the ants. A large colony of black ants was situated by the edge of the creek at a spot directly across from a colony of red ants. Jim had invited Becky to see the red ants coming over to the colony of black ants and stealing their eggs. Jim felt sorry for the black ants who were losing their offspring to become slaves of the reds. He abandoned the role of impartial observer and took a stick to stir up the sand between the two colonies and drove the red ants away. They took no black ant egg hostage while he was around.

Do You See What I See?

Becky H. had a playhouse in the Willow tree near the creek. At sometime in the past a roll of old woven wire fence had come to rest near the tree. Willow branches growing into the wire kept the creek from carrying it away. The roll of woven wire became a good "home-grown" trampoline. Becky's brother David managed jumping barefoot on the wire fence trampoline.

One of the fringe benefits at the end of school each year for David H. was the permission given by his parents to go barefoot. Children walking around with shoes untied are simply witnessing to the primordial cry to free the feet. Once tender feet are again tough as toe nails, no matter how often toes are stubbed it is worth the short-lived pain to bare your sole to the soil again.

Donald G.'s parents left it up to Mother Nature to decree when Donald could set his shoes aside. 60 degrees was the bench mark. He kept an eye on the thermometer every day in late spring. There were no variables for me. It was useless to plead for a variance. May 1st was the day even if it was 110 degrees in the shade on April 30th. That obviously did not happen.

See what you miss

For those who have not had the pleasure of watching dung beetles, "tumble bugs" in action it may not seem to be upscale entertainment. Jim, Becky and David H. recalled times of fascinating observation of this barnyard intrigue. Just as squirrels carry walnuts for a place of storage, dung beetles take their prize of captured dung to a place of hiding. The beetles do not have the advantage of size which makes them all the more fun to watch. The harvest of waste from farm animal deposits goes by various names. Dung is one of them; manure another and I'll let it drop with that. Dung beetles are about an inch long and they harvest the dung from cattle droppings by forming it into a ball about their size, an inch in diameter. They seem to be equally adept at rolling or pushing the one inch ball of manure with their front feet on the ball and rear feet on the ground, or the other way around when the need arises. All goes well if the ground is flat. However, if there is a bit of a hill the ball picks up speed and rolls the beetle. This calls for the help of another beetle since it usually takes two to roll the dung ball back up the hill. They have no laborsaving devices at their disposal. With what seems to be simple strength and awkwardness the beetles get it done. For children this was a "hey, look at this" no cost event.

What's going on in the barn?

It seemed like a foolish waste of my time to look in the dictionary for a definition of the word barn. Even though I was born in a hospital in Omaha, Nebraska and not in a barn, I spent many hours in and around a barn on our farm in Illinois. Foolish or not, I found in the dictionary, "A barn is a building for storing hay, stabling livestock, etc." Had they asked me, I could have added a half page more between livestock and etc. They didn't ask and I won't tell them. In fact I will not insult anyone with a long list. At this point all I'd like to add is one thing; a barn is used for recreation.

I have fond memories of a Halloween party held in our three story dairy barn. Authentic rural decorations were no problem with corn shocks and pumpkins. The genuine cob webs were spun by local spiders that were barn born and bred. Resident black cats made their appearance from time-to- time. Hay and straw bales were both decorative and functional as tables and padded furniture. Surprisingly real noises came from surprisingly real cows installed in stanchions on the ground level. From time-to-time the hand cranked corn-sheller was wound up to a top whine. The "two banger" John Deere tractor was started up when things got dull. A wind slamming the barn door on the second floor was an unscheduled gift from the spirits of the barn. A guided tour took visitors up the vertical ladder to the haymow and down the back stairs to the ground floor. Here among the infestation of animals, the entertainment tour was concluded with a ride on the overturned, round bottom, semi-clean, manure carrier tub. The tub ran on a track behind the cow stalls round a sharp curve and over the manure spreader at the back of the barn. Our city guests, two or three at a time, were in for a close up view of the rear end of a line of cows on an exciting, noisy trolley ride that ended over the half-filled manure spreader. Then it was time for the box lunch meal. Barns can be a place to go for fun. Don't you agree?

Before hay was put in their haymow with a smooth newly finished floor, Mary Lib S. and family decided the time was right for a square dance. Friends and neighbors were invited to a carry-in-meal to be followed by a square dance. All agreed it was a great way to dedicate a new barn floor soon to be filled with the harvest of hay.

The haymow is a good place to play "hide and seek." Burdette Q. and other young friends found that to be true. While there are many dark corners around stacked loose hay and corn fodder, bales of hay and straw lend themselves for making excellent hiding places. Bales can be moved and stacked is such a way as to create tunnels and voids making it hard to be found. In "hide and seek," if you like the pretty girl that's doing the seeking, you don't want to be too hard to find. Some of us bale movers know the game's over if you forget to put them back. Dads are not fond of falling into the void of a bale tunnel. A stack of baled hay full of tunnels and voids can be misleading. When dad finds out there are not as many bales at it appears, he may be just a little upset with you. It was also smart to put them back while you had others to help.

Burdette and friends found the haymow floor a good place to play basketball. During the daylight hours it was great fun, but the boys had more time to play in the evening when they had finished their chores. This was not the best time because of darkness, but they hung up a kerosene lantern. Some basketball shots went in because of a definite lantern court advantage. Shots at the other side of the lantern light didn't fall in quite as often. Every shot was in trouble when a wild wind went through the barn causing the lantern flame to flicker. It was a good excuse, or reason for missing the shot. These happenings on the hardwood of the haymow were all in good fun even without referees, penalties and a roaring crowd.

On getting a wagon ride the hard way

Why was Eleanor G. pulling the little red Radio Flyer wagon to Dover Friends Meeting on a sunny Sunday morning? Why was her daughter Becky riding in the wagon? This is the story. Eleanor had gone to help her husband Phil drive the beef cattle into the barn. She could not leave her young daughter Becky home alone, so she brought her daughter with her to the cattle barn. Here Eleanor left Becky on a stack of panels in what she thought was a safe place away from the cattle. As soon as they were out of sight and working getting the cattle situated, Becky decided she wanted to get a better view of the action and climbed up higher on some wooden panels. Becky was pleased with her climb until it happened. She had no sooner reached her goal when the panels suddenly gave way. Both Eleanor and Phil heard their daughter's scream of fright and pain. They couldn't be certain, but it looked to them to be a broken leg. Their local doctor examined the break. It was worse than they thought and required a trip to Children's Hospital in Cincinnati. Becky was in a body cast for many weeks while the bones grew back in place. When she was finally able to be out and about and on her own, Becky had a story to tell of her cast and the rides in the little red Radio Flyer wagon.

Winter has its down hill slide

Mike D. had a new idea sneak up on the blind side of his reasoning. He did not pay it much attention until the thought was comfortably seated inside the closed door of the fun loving section of his brain. Then, and only then, did it call out to be recognized. Mike and his two sisters loved the down hill sledding at the Snow Hill Golf Course. When the new idea mentioned a speedier way to zip down the snow covered hills, it had Mike's full attention. That's it, he agreed. A sled with two thin runners that sank down in the snow could not go as fast as one with a wide surface. Such a sled would not sink into the snow. Mike did not have this sort of sled and wondered what might be used instead.

Once again the new idea was up front in his mind saying, "What about the door of an old refrigerator? Take a look at the 'beyond repair' refrigerators in your family's hardware store." Without taking note of the comments of caution counter to the new idea, as youth often do, Mike went into action to make a super sled. After he located an old refrigerator without any hope of happy humming in the future, Mike took off the door and ripped out the liner of insulation. In its place he put a blanket for comfort seating. He knew it was going to be fast.

At the hill of their choice, Mike put his two sisters in his sled and gave them a shove. Away they went. As he'd hoped, the newly created sled zoomed down the hill at record speed. At the bottom of the slope was a flat stretch where the old sled usually slowed to a safe stop. That was good because beyond the flat was a sand trap and then the boundary fence. In this case the natural breaks were of little help. The thrill was compounded into fear as the girls raced down the hill, sailed into and out of the sand trap and landed in a heap against the fence. Mike rushed down and discovered the girls were not hurt, but they did not want to do it again. Mike agreed with his sisters. Another thought whispered in his head suggesting he should give it a try. However, he rejected the idea of trying a slick slide on his super sled. Mike was not afraid to try. He simply had no desire to repair damage to the fence.

Safe sledding for children—no cars allowed

As a rule, it is not wise or safe to take a sled for a slide on a city street or even a less traveled country road. Jean H. was once in violation of this rule of common sense for the sake of safety. Sledding down the hill on the road in front of their farm this day was great fun for Jean and her friends. At their end of the road there is a long north sloping hill where a kid on a sled can go on and on and on, almost forever. At least it seemed that way when they had to drag their trailing sleds back up that long hill. This was a special day for sledding on the hill because they were doing so with the approval of their parents and the local authorities. Prior to this time of fun for the kids, adults in their cars were unable to make it up this slippery slope. Road crews were so busy with clearing other roads; they had not found time to get to this hill. Adults were not too happy to see the ROAD CLOSED sign at the bottom of the road. Rules of no sledding were set aside by the sign. No one got hurt. This was Children's Day on the hill right in front of Jean's house. Did not Christ say something about letting the little children come unto Him?

On making a temporary indoor ice skating rink

At a neighborhood get-together, in Illinois on a cold winter night, the adults were in one part of the house playing cards. The young people were in the large farm kitchen. The kitchen door usually open to the rest of the house was closed. Both the adults and children were comfortable with this arrangement. Privacy was the one point of agreement for those gathered on each side of the door for this night. One of the youngsters came up with a "neat" idea for some fun that was countered by a "double dare you." ("Neat" and "double dare you,") were the overworked words of teens in the 1940s; much as "You know" is today, you know). The "neat" idea began by creating a place to skate by pouring a little water on the kitchen floor. Ice for skating was made possible by opening the windows and the door to the winter's fast freeze. It was all great fun on a slippery slide until the adults began to wonder why it was getting so cold in the house.

Winter's fun for flatlanders

If you were unfortunate to live where there were no sledding hills nearby, you could always do as Sharon G. and make snow angels. While it may not be listed as one of the most popular winter sports in the snowbelt, the tradition and artistry have not completely melted away. It is one of those rare cost-free things to do for fun. All you need is a few inches of snow on the ground. This usually comes at no charge to children. It is advisable to have a warm water-proof coat, hat, gloves and boots. The actual creation of a snow angel requires little effort or long range planning. Standing on fresh snow covered ground with legs spread apart and outstretched arms simply fall on your back. (Believe me at 80+ years, that's not hard at all. The challenge for me is finding someone who is willing to help me get back up). Now where was I before that rude self-interruption? The next step in the process is to rake the loose snow moving your arms up and down. If you can sit up and get up, you will leave the imprint of a winged snow angel. A final word of caution, if you have a parent, or parents who have the mistaken notion that you will "catch your death of cold," (whatever that means), playing on the ground in snow, here is my suggestion. Show them your heavenly artistic creation before you show them how it's done. The only hitch to the picture of snow play is the probability they did the same thing when they were young. If that is the case, challenge them to join you. Have fun.

The key to skating

Sharon G. also remembers ice skating on their frozen pond. You can't do that very well on a sledding hill. If you were a beginner, chances are you had clamp on skates fastened to your shoes with clamps tightened by a skate key. Unless you were good at putting things away where they belong, skate keys often hide in the strangest places. There was not much use in asking your mother, "where is my skate key?" because she would always ask, "Where were you when you took off your skates?" How would you know? Skate keys often were the same size. If you had a friend with one in his/her pocket, that saved a lot of looking.

A frozen "crick" does the trick

The creek that ran between their house and the barn also ran some distance to a neighbor's pond. In the winter time when the creek was frozen Dick G. could put on his ice skates, and skate on the frozen creek all the way to the house and pond. He could, but why would he do such a thing? Even on the coldest of winter days, you might find Dick skating in the direction of the pond. He enjoyed skating. Did I fail to mention there was a fair young maiden about his age who lived in the house beside the pond? I probably did forget to mention it because the frozen creek on our farm in Illinois ran no where near a fair young lady or a frozen pond.

Free snow—take all you want

In 1936 Dotty M. and her family lived on Katy's Lane near Wilmington, Ohio. Big snow storms made frequent visits that winter after a very dry summer and fall. One such storm was followed by freezing rain. Heavy coats of ice on top of the snow made it very difficult to clear the driveways and roads. Those days saw no snow plows or heavy machinery in rural areas to clear the ice-bound snow from the roads. Farmers and neighbors got together with shovels and picks and cleared the roads. Because of the combination of snow and ice, the removal resulted in big chunks of ice-bound snow. The simplest way of disposal was to throw them into the side ditches.

Do You See What I See?

Working men and older boys saw nothing more in these chunks than blocks of obstruction to travel on the road. Dotty and her brother saw them with the light of imaginative eyes. These cast-off chunks were ready-made building blocks. With efforts known to children with excess energy born of a creative idea, they built an igloo. No one told them, "Leave those ice chunks alone," nor did they say, "You kids get busy and move these ice chunks away." No money came from their pockets for this project. With the currency of their imagination and dreams, Dotty and her brother invested their own energy in the project. In return they received the rich reward of hours of enjoyment and satisfaction in their own hand-made igloo. When the sun's warm rays returned the ice to the water thirsty soil, the children were ready to move on in their understanding God's way with the world. There is cause for hope and rejoicing, when our children discover something of worth out of that which was cast aside as of no account.

When new won't do

In the mid 1940s when times were tough weather-wise and financially, Press A. and his wife Katy agonized over what to get their two children Stew and Sara for Christmas. It couldn't be much, but they had to give them a little something beside their love and care. With only a few days left before Christmas, they were able to discuss the matter openly, since the children were outside playing in the snow. Winter came early that fall on the farm in central Indiana with a promise to stay for some time. After while, Press stood up to look out the window to check on the children. What he saw brought the answer to their question of what to get them. Stew and Sara each sitting in a circular, metal hog feeding pan, were sliding down a hill near the barn. "I know what we can get them," Press told his wife. "Those kids need a decent sled!" Katy thought it was a great idea. The challenge for the next few days was to search the stores in town for a sled big enough for what little money they had to spend.

On Christmas morning the sled of choice made of sturdy wood with shinny steel runners stood beside their Christmas tree. The children had helped their father cut the tree and drag it through the snow from the nearby woods. Their mother helped them with the home-made decorations, including a long string of popcorn. When Press had finished the early morning chores of caring for the livestock, he returned to the warmth of the house and the excited children waiting impatiently in their bedrooms for that call of "Merry Christmas." Press and Katy were amply rewarded by the joyous response of their children. Stew and Sara thought they couldn't wait to go out and play with their new sled. However, their mother insisted they have a good breakfast and be warmly dressed before they ventured out to the hill with their pride of Christmas, a brand new sled.

After the parents saw them safely situated on the hill with the sled, they sat down at the kitchen table for another cup of coffee and a time of counting their blessings. Later Katy went to the window and what she saw prompted a spontaneous outburst of laughter and "Oh Press, come and see the kids!" What did Press see that was so funny to Katy? The new sled was standing up against the fence. Stew and Sara were sliding down the hill laughing and turning around and around in the old hog pans.

Free homemade toys

The well worn word "free," is one of the most deceptive words of our time. I am told I get "free" alignment and rotation of the new tires I purchased. The food my wife buys for our meals, is sugar and fat "free." My doctor hands out "free" pills furnished by some pharmaceutical company. Most of us know these claims are false. Somebody has to pay. I think I know that somebody.

Many toys we had on the farm did not involve

monetary exchange, but they took time and labor. The greatest gratification for older children came from the labor of making their own toys. Homemade toys were secured by the coin of imagination and time spent foraging around stock piles of discarded junk. It often took hours of construction trial and error. When all else failed, children could get the help of adults. The latter was usually only as a last resort after a little blood, sweat and tears. Gauze, tape, Iodine and dad's bandana or mother's wash cloth figured into many a homemade toy.

    Virginia B. had the help of her grandpa to make whistles out of a branch cut from a Papaw tree. A Willow branch worked just as well for others. These instruments of nature's fine notes or noise were about 3 inches long and ½ inch in diameter. Whistles were crafted from young branches usually in the spring when the bark could be slid off without breaking. An air chamber for air passage was hollowed out of the wood. The bark was then slid back in place with a slot cut into it as the blow hole for the whistle. Warning, use a sharp knife carefully and consult with a grandpa who knows how to do it.

Getting a kick out of butchering

    Wesley G. is old enough to know far more than a person can discover at a meat counter about butchering a hog. There is more truth to the old saying of "using all the hog parts but the squeal" than most of us know today. Can you imagine a part of a butchered hog for a ball? I find no reason to doubt Wesley's claim of using the bladder for a ball. As he recalls a hog bladder was cleaned, tied at one end and then the other end after it was inflated with air. Did you not know, have you not heard the football called the pig skin? Kick that around in your brain for awhile. Waste not, want not; whole hog or none. Homegrown and homemade toys are free.

Take time to whittle

Gene T. tells of toys made at home in Breathitt County, Kentucky. When you didn't have money to buy toys, you made them. One of the basic tools for entertainment and toy making was a pocket knife. A boy was blessed to own one given to him by a grandfather or a friend. Much as golf for some today, whittling had almost no age limits. Even though one could just sit and whittle for pleasure and passing the time, a good sharp knife and the right piece of wood was toy material.

The ring is the thing

For entertainment they also played Horseshoes. I'm not putting down the present popular game of Corn Hole, but it just doesn't measure up to the triumphant ring of a genuine horseshoe when it finds its mark and rings the stake. Gene told us they often used mule shoes where he grew up. Shoes for mules are longer and not as wide. It seems to me using mule shoes would make it a bit more of a challenge to ring the stake, but the shoe would come nearer sticking with it. If they are still playing Horseshoes with mule shoes, I suggest they quit "horsing around and bite the bit." Call the game the tougher name it deserves. As I sit here, a senior citizen slumped over at the computer key board, it's easy for me to say, "Make mine Muleshoes!"

Use for play—then recycle

I am a fan of recycling cardboard. Nearly every week you may see me dragging out boxes that have been shoved into the dumpster at our rental apartments. Cardboard boxes can be put to better use traveling the recycling route. To be recycled and used again is far better for cardboard than riding in the trash truck to the landfill to waste away among rotten garbage.

Dorothy D. reminds us of a creative use of cardboard boxes after they are too old and worn to be a trusted carrier of goods and before flattened for recycling. Dorothy found big old cardboard boxes and gave them to her children and their friends. "What could you make with these big boxes?" she asked. When one of them managed to crawl inside one of the boxes, others decided it could be made into a playhouse. With Dorothy's help furnishing crayons, pencils and safe cardboard cutting tools, the children were set for hours of fun. They cut holes in the boxes for windows and doors. Crayons were used to make flowers, trees and other scenery. Let us not forget children can use their imagination to entertain themselves with inexpensive toys, often of their own making.

Not unlike cardboard with greater potential than disposal as trash, our children gain a better understanding of their creativity and self worth. When the labor of learning is fun, our children are blessed with a glimpse of what labor as an adult can and ought to be. There is hope for the future, if they remember instruments of labor and resources of the good life need not be tied to the latest and most expensive devices.

Pets and play

Two boys might be able to ride a horse in the saddle, but what do you do when four boys want to ride at the same time? The answer was obvious to Burdette Q. and his friends. Take off the saddle and ride the horse bareback. Older horses on the farm usually were not subject to flight. This horse didn't seem to mind once the four boys all managed to make it to the top. Not one of the boys tried the rear mount trick like the western stars of the movies. The boys managed the horse mount with the help of a climb on a fence gate. Having made it to the top away they rode. Burdette was up front with each boy holding on to the one in front of him. It was all a grand time of laughing and bragging of their bareback riding skills.

As one might guess, the whole hilarious spectacle had its inevitable downside. Burdette, without a secure hold, was the first to slip off the horse. Head first, he hit the ground with a thud. The other three were quick to join him predictably falling to the ground.

Actually they fell on Burdette. In the light of day, Burdette saw stars that his friends did not share. The old horse seemed unfazed as if this happened all the time. In time Burdette recovered enough to tell the tale and enjoy the humor of their bareback ride.

A not so lucky buck

Most farm folk would understand when you talked about the "buck" in reference to sheep, you were referring to a male in the flock. In my dictionary "buck" means "an attempt to dislodge a rider or burden." It also can mean "to butt with the head." A male sheep is often aggressive and is labeled the "Buck sheep" for obvious reasons. Everyone who's had any experience with the male of the species knows something of the surprise of a buck attack when your back is turned. For some young people, mostly boys, it becomes something of a game. It is called, "Teasing the Buck sheep," better played with the help of another boy or two to watch when your back is turned. We always felt buck sheep were so named because of their head butt. Turn your back to the buck and you were inviting an attempted face down grounding. It probably was justified by the tormenting my brothers and I gave him. We were testing the rumor that a buck sheep closed his eyes when he lowered his head in an attempt to get you. We tried to step aside the moment he lowered his head so he'd have a near miss. Sometimes it worked, sometimes he got us. It was all part of recreation on the farm.

Don't tell him he is not a pony

Susanne K. had a Southdown buck sheep that may or may not have been inclined to buck. I don't think she was into the sport of "buck-bating." From what she told us we concluded their buck did not buck her. Susanne put a pony saddle on their Southdown buck and rode him around in the barnlot. That, I'd like to have seen! Do you suppose Susanne ever let him know he was not a pony?

Do You See What I See?

A run-of rabbits

Rick K. raised domestic rabbits as a hobby. He sectioned off one end of the chicken house with an enclosure for each doe and litter. In the rabbit section of the chicken house Rick had seven litters at the same time. These rabbits were not sold as pets or for food. Rick enjoyed watching the rabbits. They were Rick's pets. Dare I say prolific pets? Eventually they were turned loose. The farmstead was blessed with rabbits, rabbits, rabbits running around all over. In winter when it was cold they would stay in the cow-shed under the manger. What more could they want? Here the rabbits had warmth, water and hay all winter. Even though the rabbits were not sold or slaughtered for human consumption, word got out of their existence among the animal predators. "Survival of the fittest," is something more than food for thought.

On getting a kick out of riding

As a youngster Sharon G. had a bit of bad luck with Midgie, their little old pony. Midgie was old, tired and set in her ways. To be more specific at the time of this encounter the old pony was in the way. Sharon decided to help her move by whacking the pony on the rump. Even though old and tired, this got Midgie's attention. She made her position on the matter very clear to Sharon by kicking her in the stomach. Sharon did not try that again. With an old somewhat lifeless pony that was not too much fun to ride, Sharon and friends considered the family's dairy calves. The calves were about the size of the pony, so why not give riding them a try? The dairy calves took a firm stand in opposition to being satisfactory substitutes for a pony. All attempts to ride them fell flat. The story had a happy ending for both the kids and the calves. At the age of seven, Sharon got a Shetland pony named Rocky. The pony served her well until she was in high school. By this time Sharon was old enough to handle a horse. Sharon acquired a horse named Betsy. The "kick" she got out of this horse was heartfelt. Betsy afforded Sharon many happy hours riding.

Napoleon's retreat

Becky H. had a pet wether (neutered male) sheep she named Dorsey. In their flock of sheep they had a buck sheep named Napoleon that lived up to his name. When Becky had the task of putting hay out for the sheep, she always kept a wary eye on Napoleon. He seemed to enjoy bucking people more than eating the hay. Once on a memorable day, Becky successfully escaped the charge of Napoleon by climbing up into the rack that held the hay. While Napoleon was standing by the rack daring Becky to come down on the level fighting field, the buck failed to take notice of her pet sheep Dorsey. Had Napoleon turned to meet Dorsey's charge head on it might have been a good scrap, but Dorsey blind-sided the buck and sent him rolling. Napoleon retreated still shaken-up wondering what hit him. With her pet Dorsey now in charge, Becky was free to climb down from her safe perch and finish her task.

Becky's spin on the story ends with praise for Dorsey and the charge responsible for Napoleon's retreat. She feels her pet sheep had the courage to protect her out of gratitude for her special care and friendship. Brother Jim's comment was, "Dorsey was just hungry for the hay."

Potpouri of play

This is far from a complete list of recreational toys, games and activities of the youth on family farms and in small towns in the 1930s and 1940s. It is simply a few of the activities remembered by those interviewed for this book. The games and activities brought back memories of a time when recreation's focus was on participation, personal and group, with little or no financial outlay. Today in sharp contrast recreation often involves a substantial financial outlay with limited participation. All too often now we pay dearly to watch others play.

The right way—homemade

Bows and arrows made from tree limbs and shingles, sling shots made with a forked stick and inner tube rubber, whistles made from Willow or Papaw trees, shooters made from tire inner-tube rubber and scrap wood. Paper dolls made from catalogues.

The right price—or trade

Every boy had a pocket knife, most had a sack of marbles, some kept the marbles in a small bag that used to hold tobacco (when men rolled their own cigarettes), dominos, checkers, carom, soft ball, croquet, jumping ropes, girls liked to play jacks, making corn-shuck dolls, and ping pong.

The right place—to do it

Catching lightening bugs and putting them in a canning jar, watching and catching birds, bugs, frogs, toads and other creatures, wading and swimming in a creek or a favorite swimming hole, see saw/teeter totter, fishing, climbing trees and playing in the woods, making and playing in bale tunnels in the haymow, jumping down onto a pile of hay on the floor or into a wagon of grain in the barn, touch football, and coon hunting.

The right people—to be with

Playing Fox and Geese in the snow, walking around town on Saturday nights, ball games at school, Red Rover, seeing shapes in clouds, cards, Old Maid, Flinch, Pit, and many more.

We may run out of money, but it cost no cents and makes no sense to pay to see others play when recreation is free for all.

Now You See It—Now You Don't
Changes: People, Places, and Equipment

The ideal designs separated the farm's work from the home by using trees and shrubs to hide the stable, yard and outbuildings from the home. Separating work from the home paralleled urban ideals of the home as a sanctuary from the work world and stressed the division between the male and female spheres. The modern farm no longer served as the site of agricultural production, but demonstrated class status and increased consumption altering rural patterns of work and socializing.
Mary Neth—*Preserving the Family Farm*

Often in my mind's eye I visit the place of my youth on a 300 acre farm in eastern Illinois. The few times that I have actually returned I am confronted by the separation of what used to be and what is now. The youth of my past asks, "Where is the house that was our home? Where are the barns and sheds? What happened to the old orchard? Where are the fence rows and the Osage Orange trees of the hedge? And the cow pasture, the grade school, the church, where are they? The old man of my present speaks to his youth. "Look away and close your eyes. You will see they are still there."

The self-sufficiency of the family farm was challenged by the entrepreneurs who brought town and city merchandise and impulse buying to the farm families instead of waiting for them to come into town. Laborsaving equipment on the farm and in the home resulted in more independence and less need for labor and equipment sharing. The price of improvements meant a greater need for money and financial arrangements. As more laborsaving equipment was added, so was the need for more money. For those without major savings it meant an increased expense to pay off a bank loan. The answer seemed to be in a greater investment in livestock numbers and acres of land. Farm diversity took a hit when there was more money to be made in grain than livestock. Mega-farms without livestock replaced labor with larger machines for planting and harvesting grain. This added to the continuing decline in the number of people living and working on the farm. The greatest loss to rural communities is not monetary. Our greatest loss is in the satisfaction of working together and being neighbors with a shared appreciation for land and livestock.

A way, or away with the pitchfork

In our basement we have a treasured, work-worn, sweat-producing, muscle building four-tine pitchfork. The fork is treasured because it was one of the basic manual labor tools of the family farm in the 1930s. I remember such a pitchfork with a strong tie to the table fork and a hearty meal. Perhaps the single greatest change in the use of the pitchfork came with the perfection of the hydraulic cylinder. With the advent of the computer and wonders such as unmanned machines to gather moon rocks, I look at the pitchfork and ask, "Could this, and so much more have happened in my lifetime?"

An age-old cartoon characterization of the devil shows him with an upturned pitchfork in hand. Consider as well the "American Gothic" painting of Grant Wood, also with upturned pitchfork. I am no judge of art, but I think there is something about the pitchfork reflected in the absence of a smile on the face of both the man and woman.

From what he told me, Bob Mc. didn't smile much at the time of a pitchfork experience working in his uncle's horse barn. With sweat in his eyes dripping off his brow, Bob must have had a "devil of a time." His uncle used corn stock fodder for bedding in the horse barn and saw no reason to be in a hurry to clean out the bedding and manure. As long as it didn't get built up so high the horses couldn't go in and out of the barn, there was no problem. Bob's uncle knew his horses were smart enough to duck to keep from hitting their heads on the top of the door going in and out of the barn. When it was obviously time to clean the barn for the horses, you can guess who got the job. With pitchfork in hand, Bob struggled to break up the horse trampled bedding mass made of a mixture of manure and corn stalks. Bob admits the pitchfork was the tool of choice at the time. Can you imagine trying to work at it with a garden hoe or a coal shovel?

Even though I was not old enough to work in the hayfield, I shared my father's satisfaction seeing the local International Implement dealer coming down the lane to our farm with a flat-bed truck loaded with two new laborsaving implements for harvesting hay. I had seen pictures of a hay loader and a side-deliver rake, but never thought I'd see one on our farm.

Prior to the purchase of these two new implements hay was cut by a sickle-bar mower. The loose hay that dried in the field was prepared for pickup by a horse drawn sulky hay rake, (modified to be pulled by a tractor). Hay raked by the sulky rake left the hay in bunches all over the field. The bunches were loaded on the flat hay wagon by two men with pitchforks plus a man on the wagon to stack the loose hay to keep it from falling off on the ride back to the barn haymow.

The new side-deliver rake put the hay into windrows that were picked up by the hay loader pulled behind the wagon. One man was still needed to stack the hay on the wagon, but the two men loading the wagon could take their pitchforks and look somewhere else for work. These time and laborsaving improvements went well until our neighbor bought a hay baler.

The baler of square-end bales was a three man/boy operation. In addition to the person who drove the tractor pulling the baler, two were needed to sit on the platforms situated on either side of the protruding bales. Their job was to wire tie the bales before they were pushed out to fall on the ground. In many instances this noisy, dusty, sweat-producing task was turned over to older boys and in some cases girls. Custom bailing was "good money" for them, especially those who were able to work most of the summer months in the hay going from farm-to-farm for farmers without a baler.

With the progression of improvements, work on the baler ceased with the new one-man balers that tied the bales automatically with baler twine. Young people had to look elsewhere for gainful employment. Just when many thought the automatic tie baler was here to stay, innovational farm equipment engineers rolled out the more efficient round balers. Now even the stoutest farmers need mechanical help to lift the tightly packed heavy cylinders of hay. No more back breaking bucking the 60-75 pound bales. Do we bring back round barns for round bales? No, these 1000 lbs. or more are machine weatherized with plastic wrap and left lined up at the edge of the field until needed later for cattle feed. Sooner or later a tractor or truck armed with a long Unicorn-like pointed pole will arrive to spear one of the heavy round bales. With mechanical muscling, the bale will be carried without much manual labor to the place of livestock feeding. In less than a century, the sulky rake, hay loader, side-delivery rake, wire-tie baler and the automated twine-tie baler have been superseded. Now you see it—Now you don't.

The following list of contrasting changes in the home and on the farm is by no means a detailed listing of the multitude of articles of change. It is more like a brisk walk by the exhibits in a museum without taking time to examine articles or read the informative signs. My intent is to simply highlight the contrast of basic changes for people, places and things in place in the 1930s and today. The changes are almost beyond belief.

Changes in farming equipment
From:
- the walk-behind, horse-drawn, single shear plow to "no-till" machines,
- the horse-drawn one row seed planter to the tractor drawn 32 row computerized machine,
- the horse drawn McCormick reaper and thresher/separator to a computerized combine without a man on the machine,
- hands on cow milking to a computerized no-hands machine,
- ten gallon milk cans to bulk tank operation,
- the scoop shovel to grain augers and elevators,
- the hoe and scythe for cultivation to chemicals and genetic seeds,
- hand squeezing sheep shears to electric shears,
- barns for horses, cattle, swine, sheep and poultry to grain equipment and storage.

Changes in crops
From:
- open pollinated to genetically altered corn and soybean seed, etc.
- crop rotation to chemically enhanced soil,
- 2-4 stalks of corn 36 inches apart in 40 inch rows to 4 to 6 inches apart in 30 inch rows,
- 45 bushels of corn per acre to 245 bushels per acre,
- a variety of grains and grass to mainly corn and beans.

Do You See What I See?

Changes in livestock

From:
- 118,000 hogs and pigs in Clinton County (1955) to 18,376 (2007).
- 28,500 cattle and calves in Clinton County (1954) to 2,001 (2007).
- 63,400 sheep and lambs in Clinton County (1955) to 1,281 (2007).
- 364,619 hens in Clinton County (1929) to 1,110 (2007).

Changes in the home

From:
- scrub board, tub and ringer washers to multiple-choice setting automated washers,
- clothes lines in house and outside to multiple-choice settings automated dryers,
- dish pan to multiple-choice settings automated dish washers,
- wood burning cook stove to microwave ovens,
- wood and coal fired heating stoves to multiple heating options,
- candles and kerosene to LED lighting
- outhouses and WPA toilets to indoor plumbing with warm water and a shower.

Changes in the community

From:
- neighborhood merchant's credit with no interest or late fees to credit cards excesses,
- weekly community band concert to personal MP3 players.
- 42 churches, 5,368 members in 1930 to 26 churches, 1,640 members today—(Quaker),
- 76 one room schools 1918 to 8 grade schools and 4 high schools in Clinton County, Ohio.

Saved Labor—Lost The Farm
What Are The Grounds For Hope?

"Emerging doubts about unbridled individualism, consumerism, and unsustainable growth echo a disquiet found in many sectors of American life.... It is through these doubts and anxieties, this sense of loss of agrarian values and limits to industrial values that a different future may emerge... Writers who dismiss the national homage to the family farm as romantic nostalgia for a simpler, pre-industrial past fail to acknowledge its roots in frustration with the industrial life."

Peggy F. Bartlett—*American Dreams: Rural Realities*

Sometime in my preteen years after a halfhearted acceptance of my chores, I kept thinking of labor-saving ways to get it done. Feeding the chickens at times involved carrying heavy buckets of feed to the chicken house from the barn. The barn, about a hundred yards away from the chicken house, seemed such a long way to carry feed. I spent many valuable play hours working on a cable system for transporting buckets of feed from a high point in the barn to a lower point in the chicken house. For some reason Dad insisted on my feeding the chickens before messing with my laborsaving miracle cable that literally never got "off the ground."

Laborsaving dreams back then were of a robot in human form that would labor at our bidding. Nightmares of robots out of control or taking over the world stirred the imagination. Little did I expect to see these laborsaving robots become a reality in my lifetime. Nor did I anticipate the tremendous variety in shape, size and function of labor and life saving robots in operation today.

The first laborsaving robot that made a huge difference for me on the farm took the form of hydraulic power. What an equalizer it was for me, a lightweight teenager cultivating corn! The heavy gang of cultivating shovels mounted on each side of our tractor took all the manual power my weak arms could muster to raise them up out of the ground at the end of the row. With a new tractor and hydraulic power, I could lift the shovels, turn the corner and drop the shovels back down to cultivate another row with the best of stronger adults. Little did I think about it at the time, but in a small way I was a party to these troubled times when robots have taken over. As a young boy, I helped put a man out of work.

In the early 1940s my father borrowed money to purchase a combine. I doubt if Dad gave much thought to the long-range effects this laborsaving robot had on the family farm. Burdette Q. like my father had no long-range view of the consequence of the coming of the combine. As he reflects on it now it is clear with the coming of the combine, one man did away with the whole threshing crew. With the combine came the most dramatic labor layoff. In the space of a few years, perhaps a decade, one man replaced the labors of as many as two or three dozen men and women who participated in one threshing ring.

Take a look again at the article written by Dean Houghton. (See p.117). Do you see what I see? Fifty men and one hundred horses out of work for every modern combine. As one who had far more experience on the tractor side of the divide between horse and tractor power, I enjoyed working with our tractor and combine. In spite of the dust and noise, I was happy hearing the hum of the machine that separated the grain from the chaff. The combine is only one of a multitude of laborsaving robots involved in the fast fading family farm. I highlight the combine because it was so dramatic in the difference it made to rural life. In a matter of only a few years many farmers owned a combine. They no longer had need of threshing rings. This abrupt abandonment of the seasonal celebration of threshing had far reaching consequences for rural cooperation and the future of the family farm. The promotion and purchase of new laborsaving devices, preached the gospel of independence and the salvation of "each having one of their own." The eventual demise of many integrated family farms depended on a farm operation that got bigger and supposedly better. Bigger and better kept asking for more money to purchase equipment, seed and fertilizer. More money meant borrowing off the future. Credit had its added cost. To meet financial demands, family farmers had to specialize producing that which gave them the most profit.

Livestock production was lost to much larger operations. No livestock meant there was no need of fences. With the removal of fences came open space for the gain of grain. Where there is grain there is a combine. Not an ordinary Massey Harris Clipper, but a big, big combine with a comfort controlled computer room on top! To make my ridiculous rundown of rural history even more unbelievable, soon there will be no farmer or anyone else sitting in the upper room controlling the combine. Computer robots can milk cows and they can drive combines. In between the two are thousands of mini laborsaving robots working for all of us in and off the farm. Are we certain the gospel of the evil of labor and its robots are not taking over?

What have the laborsaving computerized devices of our time done to not only the family farm, but to every rural and urban laborer who has been replaced? How will the masses pay for expensive laborsaving devices if they have no job? It is not a nightmare which vanishes with the morning light. It is a frightening reality to the unemployed.

Do You See What I See?

I confess I do not feel comfortable writing as I have. The debt I owe to the laborsaving devices of this day I cannot pay no matter how long I live. From present day medical procedures and pharmaceuticals to cell phones and heated car seats, I have been abundantly blessed. With the aid of the computer and "spell-check" I have been able to write and publish this book. With that same computer aid, books may also go the way of party line phones.

Our humanity exposes us to the blessing of the curse. For those who are convinced labor is a curse, I have a down-to-earth example of a potential blessing of that curse. Consider the win/win situation of using a motor-free reel mower for grass. There would be:

1. No pollution of the air by gasoline exhaust fumes.
2. No excess noise.
3. Little financial outlay.
4. Mental and physical health benefits of labor.

I should put this disclaimer in small print. But I won't. It is not my intent to offend the lawn and garden tractor companies. The same could be said about parking the car and walking. I frequently use power tools even though they still make hand saws and drills. It is a bit more ridicules for me when I think about my electric toothbrush with a built in timer. How hard is it to turn to my wrist and look at my watch to tell when I've brushed for two minutes? Nobody will take me seriously. This is as far as I plan to push it.

Labor is not an evil to be avoided. It is an essential for the care and feeding of the whole person in relationship to others, the good earth and the divine creator. Labor on the diversified family farm had its challenges but they were far surpassed by the successes. As a participating member of a family farm, we had a rich variety of opportunities to labor. The self-satisfaction of laboring with others for the common good helps us face each day with renewed confidence. How can we reap the essential good works of labor-saving computerized equipment and have employment for all that will provide income for the essential products of life?

Gradually, year-by-year family farmers have added laborsaving machinery and appliances in the home as well as on the farm. Year-by-year like most of their neighbors, they participated in the gradual demise of the family farm and rural neighborhood. The purchase and employment of laborsaving devices did not gain them a greater degree of independence. On the contrary, family farms became almost totally dependent upon those who supplied farms and homes with essential fuel, energy and resources for the production of marketable products.

One simple illustration tells it all. Not unlike their city neighbors, the farm family became almost totally dependent on petro-chemicals, electrical energy and financial credit. Just a few weeks without access to one or all of these, spells disaster. My purpose is not to be a prophet of doom. These stories stem from a time when farmers had the satisfaction of working mostly on their own or with help from their neighbors for the essentials of a good life. Today we pay dearly for the independence of making it on our own. We have but a mirage of independence. We are under daily pressure to go beyond our means and turn "wants" into "needs." Farm families are greatly dependent on the market and governmental support. Farm families are now not much different than their city neighbors. They are caught up in the same cultural denial of need for community enabled by the choice of personal financial gain over relationship with near neighbors. Farm families did not have a monopoly on the spirit of independence, but basic to the essential functioning of a diversified family farm was the assumption of personal responsibility of every productive member.

How ironic that those who greatly prized their independence would become so totally dependent on laborsaving equipment and the provision of power to keep them in operation. Let us be united by the satisfaction of service and labor with love of neighbor rather than accumulation of money. Embrace the true profit of genuine relationships over financial gain.

The stories told me are of a time when we recognized our need for others; a time when we acknowledged our dual need to give and receive. Self-sufficiency was tied to the smaller community and not just the individual person or families.

Early fantasies before we were introduced to the computer were the fiction of robots taking over the world. It comes as a surprise to me how cleverly computers have taken away our rights to responsible labor. We need to remember we the people are in charge of the computer chips. They labor for us so we may have more time for the meaningful labor of relationship with and for others. We cannot expect governmental regulations to bring us back into cooperation, communication and contact with each other in community.

What now? Do we forge ahead, or go back and try to pick it up? I'm encouraged by the people who favor looking back in an effort to find the community we had rather than going ahead as if nothing had been left behind. Together in small groups, we can be reunited on a higher spiritual level above and beyond our petty differences. Our hope is in a firm faith grounded in our basic need to relate to each other. Our future depends on our love and support of each other for the greater good of the land, livestock and our commitment to the keeper of our spiritual and eternal destiny.

Words of wisdom and hope

Dick & Sharon G. are convinced all their children truly appreciate growing up on a diversified family farm. One of their daughters lives on a farm. The other three children would like to live out in the country where they could have some livestock and chickens. One daughter would like to be where she could at least raise chickens. She did very well with chickens at the county fair when she was in 4-H. They learned leadership skills growing up and in 4-H and FFA.

We are not without grounds for hope in Clinton County and the surrounding area. Outside the forgotten fences of family farms there is a groundswell of dedication to practical ways to recover some of the basic blessings of an earlier day. Under the broad umbrella of Energizing Clinton County, individuals and organizations are making a difference.

Growing up on a diversified family farm Darell F. found the essential values for a good life. It gave him both purpose and direction that has taken him into 34 states working as a livestock judge in county, district and national shows. He has a deep appreciation for 4-H and FFA. Darell has been richly rewarded by his work with the young people and their leaders. Darell's work as a livestock judge has been motivated by a love of livestock and children. It is truly satisfying to see the results and growth of both the animal and the child. All are winners if he/she takes pride in the care and grooming of the animals. This bonding does not just happen the week before the show. When a young person is in the ring competing for the showmanship award, it is easy to tell by the contact of their eyes, the way they treat their animal and the respect they show for others how much life on a diversified family farm means to them.

As with others interviewed, Darell is greatly concerned about the future of the small family farm and the diminishing numbers of livestock in our area.

The Wilmington College Department of Agriculture's farm has diversity of livestock and land use. While profitability of production is given consideration, the primary emphasis is upon hands-on instruction in care of livestock and use of farm machinery basic to family farm operation.

The Clinton County Farmers Market is a win/win enterprise for both farm producers and area residents who can enjoy fruits and vegetables picked at their full-flavor peak. The farmers get a fair price for their labors without the expense of shipping and handling by others.

Do You See What I See?

The Grow Food, Grow Hope Community Garden project is located on the campus of Wilmington College. The 20 small raised bed garden plots produced fresh food for low income families in 2009, but the greater gain will come from the young people being introduced to a wide variety of garden produced food, as well as instruction and incentive for gardening on their own.

This brief accounting is but a small glimpse of the activities and ideas in our area. We have hopes for a groundswell of support for community recovery.

It is encouraging to see the young people involved. The young people and leaders continuing in their active support of 4-H and FFA have served us well in the past. Their emerging qualities of leadership give us grounds for hope.

Junior Fair finalist—2009
By Danielle Hendershott

"Since the economic downturn many persons have been cutting back on their extra activities. This is especially true in Wilmington, with the extraction of DHL many 4-H participants have stepped back in order to save money.

In no way do I shun any of their choices however in an economical time such as the one we are in right now you have to consider going back to the basics. 4-H was created to help youth understand and learn from the upbringing of an animal or the erection of a project. Self-reliance is an underlying backbone of the 4-H history. Learning to cook, sew and raise animals are valuable assets to individuals.

Through proper representation and the unity of a community one can only hope that those who have strayed from the program will see what a powerful object it is in their life and come back. As a long-time member of 4-H I have seen the effect in which it has taken on its participants and I can see the effect which it has taken on myself, I only hope that it may continue to strive and impact the lives of other youth."

Do You See What I See?

Is it possible to keep all we have gained by modern technology and have interdependence and community? Today we are pressured to borrow off the future for commodities we don't have and perhaps don't really need. Unfortunately, we act as if cooperative sharing is no longer practical or necessary. Let us count our true blessings and take an honest inventory of our basic needs and get rid of the chaff of needless "wants" no matter who tells us otherwise.

The friends who shared their family farm stories also share a growing concern for recovery of the values basic to rural life and community. We admit that life on the family farm was not without its challenge to relationship both within families and community. Now sixty to eighty years later we are still convinced the rural community was well served by the functioning family farm. It is unclear as to the makeup of community in years to come, but we have little hope of peace and prosperity without some form of community that affords personal interaction with all members. Our grounds for hope are in the leadership of today's youth who are dedicated to a wholesome bonding with all creation. The earth in all its diversity: soil, plant, animal and human life is sustained by an unending gratitude to God. The lost will be found.

Think about it -

"We're all like sheep who've wandered off and gotten lost. We have all done our own thing, gone our own way."
Isaiah 53: 6

"Look at it this way, if a man has a hundred sheep and one of them wanders off, doesn't he leave the ninety-nine and go after the one? And if he finds it, doesn't he make far more over it than the ninety-nine who stay put? Your Father in heaven feels the same way. He doesn't want to lose even one of these simple believers." Matthew 18: 12-14

Eugene H. Peterson– *The Message*

Do You See What I See?

Treasured Tales of Life on the Family Farm

Interview Participants

Press Alexander
Monte Anderson
Virginia Bernard
Ray and June Cosler
Dorothy Daye
Mike Daye
Darell Furlong
Donald and Joanne Gillam
Becky Hackney Godfrey
Wesley Goings
Dick and Sharon Gray
Eleanor Green
David and Pat Hackney
Jim and Lois Hackney
Margaret Hadley
Mary Emma Haines
David and Sue Harris
Jeff and Teresa Harris
Mary Eleanor Harris

Jean Hartsock
Dot Hayes
Danielle Hendershott
Susanne Kenney
Rick Kendall
Scott Knight
Karen Lambert
Norma Lewis
Robert McCoy
Bob McNemar
Dorothy Mockabee
Bobby and Wilma Quigley
Burdette Quigley
Dotty Robinson
Mary Lib Stanfield
Dan Stewart
Georgiana Thomas
Gene and Catherine Thompson
Tom and Rosalee Waldren

Do You See What I See?

About The Author

Jim Ellis spent his early years on the family farm near Ridge Farm, Illinois. He attended Harrison Consolidated Grade School, Ridge Farm High School and was a member of Ridge Farm Friends Church. Jim attended Whittier College, the University of Illinois, Earlham College, and Christian Theological Seminary. His pastoral work began as a member of a student rural ministry team of the Wesley Foundation at the University of Illinois. Since 1950, he has served rural Quaker churches in Illinois, Indiana and Ohio. Jim has been actively involved in Wilmington Yearly Meeting of Friends (Church) since 1962. In addition to serving as pastor in Friends Meetings, for 21 years he was involved in housing management and meeting other needs of the elderly and handicapped.

Jim and his wife Anne, from former marriages, have nine children, seventeen grandchildren and eleven great-grandchildren. Retirement affords them an opportunity to travel visiting children in Ohio, Indiana, Kentucky, Colorado, North and South Carolina as well as relatives and friends over most of the United States. They have enjoyed canoeing, caving, and hiking. Something called "aging" has mandated much of this to memory. They find no restrictions on reading books as well as working puzzles and playing games like dominos with each other, family, and friends. Though officially retired they both are still involved in the work of the church.